8th International Conference on Fate and Effects of Pulp and Paper Mill Effluents and 10th IWA Symposium on Forest Industry Wastewater 2012

Concepcion, Chile
8-11 January 2012

ISBN: 978-1-5108-2335-8

Printed from e-media with permission by:

Curran Associates, Inc.
57 Morehouse Lane
Red Hook, NY 12571

Printed by Curran Associates, Inc. (2016)

University of Concepción
c/o Dr. Ricardo Barra
Environmental Sciences Center, University of Concepción
Barrio Universitario S/N, P.O. Box 160-C
Concepción, Chile

Phone: +56 +41-220-4013

ricbarra@udec.cl

Additional copies of this publication are available from:

Curran Associates, Inc.
57 Morehouse Lane
Red Hook, NY 12571 USA
Phone: 845-758-0400
Fax: 845-758-2633
Email: curran@proceedings.com
Web: www.proceedings.com

CHILE 2012

Concepción • January • 8 - 11

8th International Conference on Fate and Effects of Pulp and Paper Mill Effluents

IWA International Water Association

10th IWA Symposium on Forest Industry Wastewaters

www.forestwater.cl

10th IWA Symposium on Forest Industry Wastewater and 8th International Conference on the Fate and Effects of Pulp and Paper Mill Effluents

Concepción (Chile), January 8-11, 2012

Welcome to Concepción, Chile!

On behalf of IWA, SETAC Latin America and Universidad de Concepción it is our pleasure to welcome academics, governmental agencies, professionals and all related to the pulp & paper industry around the world to attend the "10th IWA Symposium on Forest Industry Wastewaters" and the "8th International Conference on the Fate and Effects of Pulp and Paper Mill Effluents".

The "10th IWA Symposium on Forest Industry Wastewaters" is the premier event for communicating advances in environmental science and technology applicable to all areas related to the forest products industry, particularly the pulp and paper industry. The emphasis of the 10th IWA Symposium will be the "Forest industry taking care of the environment".

The "8th International Conference on the Fate and Effects of Pulp and Paper Mill Effluents" is a series of conferences that were initiated in response to rising concern about the potential environmental impacts of the forest industry. These meetings were designed to communicate the rapidly evolving knowledge on this matter and allow discussion in a focused international forum. The emphasis of the 8th Conference will be "Integrating industry and ecosystem".
We sincerely hope you will enjoy the conferences and your time visiting Chile.

Dr. Ricardo Barra
Co-Chair 8th International Conference on the Fate and Effects of Pulp and Paper Mill Effluents

Dra. Gladys Vidal
Co-Chair 10th IWA Symposium on Forest Industry Wastewaters

Local Chilean Organization Committee

Gladys Vidal, University of Concepción (Co-chair)
Ricardo Barra, University of Concepción (Co-chair)
Oscar Parra, University of Concepcion
Carolina Baeza, University of Concepcion
Loreto Osorio, University of Concepcion
Guillermo Pozo, University of Concepcion
Gustavo Chiang, University of Concepcion
M. Cristina Diez, Universidad of La Frontera
Claudio Zaror, University of Concepcion
Jaime Rodríguez, University of Concepcion
Renato Quiñones, University of Concepcion
Alex Berg, University of Concepcion

National advisory Committee

Emilio Uribe, Chief Executive Officer CORMA BIO BIO
Ramiro Peralta, President of Technical Association of Pulp and Paper (ATCP)
Andrés Caamaño, Director Environmental Health and Safety, Arauco
Miguel Osses, Head of Environmental Department CORMA BIO BIO
Mario Basualto, Environment and Public Affairs CMPC Celulosa Co
Hernán Ruiz, Quality & Environment Manager Norske Skog Bio Bio
Pedro Navarrete, Environmental Ministry of BIO BIO Region
Bolívar Ruiz, Regional BIO BIO Director, Environmental Assessment Agency
Hugo Valeria, Clean Production Executive Secretary of BIO BIO Region

Scientific Committee Members

8th International Conference on the Fate and Effects of Pulp and Paper Mill Effluents

Ricardo Barra, Environmental Science Center EULA-Chile, University of Concepción
Kelly Munkittrick, Canadian Rivers Institute, University of New Brunswick, Canada
Aimo Oikari, University of Jyvaskila, Finland
Tatiana Furley, Aplysia Environmental Consultants, Brazil
Mark Hewitt, Environment Canada, Canada
Mark McMaster, Environment Canada, Canada
Rodrigo Orrego, University of Ontario, Canada
Tibor Kovacs, FPI Innovations, Canada
Deb McLatchy, Wilfrid Laurier University, Canada
Gabriela Eguren, University of la República, Uruguay
Jukka Tana, ÅF-Consult Ltd, Finland
Renato Quiñones, University of Concepcion

10th IWA Symposium on Forest Industry Wastewaters

Gladys Vidal, Environmental Science Center EULA-Chile, University of Concepción
Michael Paice, FPInnovations, Paprican Division, Canada
Claudio Mudado, Universidade de Vicosa, Brasil
Grant Allen, Toronto, Canada
Klaus Niemelä, VTT, Finland
Lazare Etiegni, Moi University, Kenia
Nickolaus Kaindl, SCA, Austria
Phil Pagoria, Weyerhaeuser Tech Center, WA, USA
Samuel Schabel, Tech U. Darmstadt, Germany
Sebastian Videla, Amec Earth & Environmental, Chile
Thomas Welander, ANOX, Sweden
Trevor Stuthridge, SCION, New Zealand
S.K. Chakrabarti, Thapar Centre for Industrial Research & Development, India
Woody Bryant, Weyeyrhaeuser, NC, USA

FORESTWATER PROGRAM ORAL PRESENTATIONS

SUNDAY, JANUARY 8	
09:00 - 14:00	Short course, `Cubo 4´ building: Microscopy and molecular biology tools for diagnosing biotreatment problem Instructors: M.Paice and J. Liu Using small fish for enviromental monitoring Instructors: K. Munkittrick and M. McMaster
16:00 - 18:00	Registration, `Cubo 4´ Building
18:00 - 19:00	Welcome reception
19:00 - 20:00	Chilean wine
MONDAY, JANUARY 9	
08:30 - 10:30	Registration, `Cubo 4´ Building
08:30 - 09:15	Open Sessions Plenary IWA. Grant Allen, Department of Chemical Engineering and Applied Chemistry University of Toronto, Canada. Where are the opportunities for our environmental technologies as the forest processing industry moves towards a biorefinery. 3
09:15 - 10:00	Plenary Fate and Effects. Mark E. McMaster, Aquatic Ecosystem Protection Research Division Environment, Canada. Investigating the causes of mill effluent effects on fish reproduction in Canada: lessons learned. 4
10:00 - 10:30	Coffee break
10:00 - 17:30	Posters display

MONDAY, JANUARY 9		
	Sessions Fate and Effects **Ecotoxicology of effluents and the receiving enviroment**	**Sessions IWA** **Life Cycle Assessment and ecological footprints (water and carbon footprint)**
10:30 - 11:00	1) B. Karrasch, H. Cid, S. Woelfl, R. Urrutia, C. Valdovinos, R. Barra, O. Parra. (Germany/Chile). Ecotoxicological impacts of pulp and paper mill effluents on carrying and assimilative capacities of plankton systems in Chilean rivers using a quantitative community to ecosystem level approach. 79	1) P. Wiegand, C. Flinders, G. Ice, D.Sleep, B. Malmberg, I. Lama, J. Schwartz, M. Mensink, D. Phenicie, W. Larson (USA). What do water profiles, water footprints, and water stewardship assessments reveal about the environmental sustainability of the forest products industry?. 11
11:00 - 11:30	2) T. Heid-Furley, J. Lombardi (Brazil). Main pulp and paper mill toxic internal effluents, its impact on WWTP and treated effluent quality. 80	2) Å. Sivard, M. Malmaeus, M. Almemark, M. Karlsson, T. Ericsson, O. Simon (Sweden). Effects of further effluent treatment stages at pulp and paper industries. 12
		Session IWA **Status of environmental issues nationally and regionally**
11:30 - 12:00	3) C. Milestone, R. Orrego, P. Scott, A.Waye, M. Kohli, B. I O'Connor, B. Smith, H. Engelhardt, D. MacLatchy, V. Trudeau, T. Kovacs, T. Heid-Furley, A. Slade, M. Servos, D. Holdway, M. Hewitt (Canada). A comparison of effluents and wood feedstocks from pulp and paper mills in Brazil, Canada and New Zealand and their ability to affect fish reproduction: Part I - chemical profiling and in vitro assessments. 81	1) D. McLaughlin (USA). An update on the development of numeric nutrient criteria in the United States: a forest industry perspective. 13
12:00 - 12:30	4) A. Llanos-Rivera, L. Castro, P. Vasquez, J. Silva, E. Bay-Schmith (Chile). Evaluation of kraft wood pulp mill secondary effluent toxicity in *Engraulis ringens* and *Strangomera bentincki* embryos utilizing a developmental toxicity test. 82	2) S. Videla, C. Muñoz (Chile). Comparative analysis of wastewater technology and river environmental impact: Kraft pulp mills in South America (Chile and Uruguay). 14
12:00 - 14:00	Lunch	
	Sessions Fate and Effects **Ecotoxicology of effluents and the receiving enviroment**	**Sessions IWA** **Internal control and system closure**
14:00 - 14:30	5) C. Flinders, W. Streblow, D. Borton, J. Beebe, D. McLaughlin, E.Schilling, P. Wiegand (USA). Aquatic Biology Research for the Forest Products Industry: An Overview of Recent Studies and Future Directions. 83	1) M. N.·Cabrera, L. Clavijo, M. Martikka, O. Dahl (Uruguay). Characterization of effluents from A-EP-DND and A-EP-D-P eucalyptus pulp bleaching sequences. 15

MONDAY, JANUARY 9		
	Sessions Fate and Effects **Reproductive effects / Endocrine disruption**	
14:30 - 15:00	1) A. Waye, J. Guerrero-Analco, A. Saleem, M. Taylor, J. Arnason, V. Trudeau (Canada). Pulp mill effluents and pulp mill feedstocks as neuroendocrine disruptors. 84	2) M. Borges, C. Mudado, J. Colodette, R. Chaves de Oliveira (Brazil). Effects of xylanase application on Eucalyptus Kraft-O Pulp. 16
15:00 - 15:30	2) R. Orrego, J. Guchardi, D. Holdway (Canada). Embryotoxicity of Chilean pulp mill effluent extracts in Japanese medaka (*Oryzias latipes*). 85	3) V. Da Re, L. Papinutti (Argentina). Decolorization of black liquor by 9 strains of lignocellulolytic basidiomycetes. 17
15:30 - 16:00	3) T. Arciszewski, K. Munkittrick (Canada). Comparing responses of sentinel fish species before and after the shutdown of a Kraft pulp mill in northern Ontario, Canada. 86	4) O. Pikka, M. Grötzner (Brazil). Water and wastewater management in modern bleached Eucalyptus Kraft pulp mill. 18
16:00 - 16:30	Coffe break	
16:30 - 17:30	Plenary 10 IWA and 8 Fate. Jukka Tana, Senior Consultant, Energy & Environment, Finland. Monitoring of the impacts of pulp mill effluents – a historical overview, challenges and lessons learned . 5	

TUESDAY, JANUARY 10		
	Sessions Fate and Effects **Effluent characterization / Toxicity Identification Evaluation**	**Sessions IWA** **Operation of treatment processes to minimize discharge and nutrient**
08:30 - 09:00	1) R. Lewis, J. Van Leeuwen, A. Averson, C. Chow, M. Jimenez (Australia). Investigation on color formation from pre and post coagulation treatment of *Pinus radiata* Sulfite pulp mill wastewater using nutrient limited aerated baffled stabilization basin treatment. 87	1) I. Horta, C. Mudado, C. Piacsek, V. Lopes (Brazil). Use granular aerobic sludge MBR (membrane bioreactor) to treat paper mill effluent. 19
09:00 - 09:30	2) P. Scott, C. Milestone, S. Smith, D. MacLatchy, M. Hewitt (Canada).Isolation and identification of ligands for the goldfish testicular androgen receptor in chemical recovery condensates from a Canadian bleached kraft pulp and paper mill. 88	2) J. Liu, M. Paice (Canada). Filamentous bacteria identification and control strategies. 20

TUESDAY, JANUARY 10		
09:30 - 10:00	3) R. Duarte, A. Duarte (Portugal). Chromophoric compounds from an Eucalyptus bleached Kraft pulp mill effluent: new insights from comprehensive two-dimensional liquid chromatography. 89	3) S. Chakrabarti, S. Gupta, Methoxy, R. Vardhan (India). Control of bulking of sludge in activated sludge process treating pulp and paper mill effluent. 21
10:00 - 10:30	Coffee break	
10:00 - 17:30	Poster display	
	Sessions Fate and Effects **Greenfield mills environmental impacts**	**Sessions IWA** **Operation of treatment processes to minimize discharge and nutrient**
10:30 - 11:00	1) S. Nothrop, J. van Leeuwen, R. Dewi, A. Everson, C. Chow, E. Plant, R. Smernik (Australia). Characterization and removal of carbon, color, phosphorus, nitrogen and odor from magnesium bisulfite pulp mill wastewater for the remediation of a large coastal shallow freshwater lake. 90	4) X. Qu, W. Gao, A. Chen, B. Liao (Canada). Effect of hydraulic retention time on the performance and microbial community in a thermophilic aerobic membrane bioreactor. 22
	Sessions Fate and Effects **Nutrient Encrichment**	
11:00 - 11:30	1) M. Davies, M. Lebo, P. Pagoria, E. Mannisto (Canada). Prediction of in-river responses to various phosphorus-reduction strategies at a Canadian Kraft pulp mill through development of a river-simulation model. 91	5) R. Cabrera-Padilla, P. Fernandes, E. Cleto (Brazil). Kinetics of aerobic thermophilic treatment of a kraft pulp bleach plant wastewater. 23
	Sessions Fate and Effects **Invertebrate / Fish responses**	
11:30 - 12:00	1) J. Parrott, M. Hewitt, T. Kovacs, D. MacLatchy, P. Martel, B. O'Connor, M.Van den Heuvel, G. Van Der Kraak, M. McMaster (Canada). Lab fish tests to assess the effects of pulp mill effluents. 92	6) T. Laukkanen, S. Toivakainen, H. Shahid, O. Dahl (Finland). Managing effluent phosphorus by optimizing the activated sludge process in ECF kraft pulp mill. 24
12:00 - 12:30	2) N. Ortiz, A. Elgueta, E. Habit (Chile). Dynamics of fish community structure from Itata river in a section with the influence of pulp mill effluent. 93	7) H. Shahid, S. Toivakainen, T. Laukkanen, O. Dahl (Finland). Applicability of biological phosphorus 25 removal process for kraft pulp mill wastewater treatment.
12:30 - 13:00	3) G. Chiang, K. Munkittrick, M. McMaster, R. Urrutia, C. Concha, R. Barra (Chile). Reproductive and metabolic responses of *Percillia gillissi* and *Trichomycterus areolatus* populations after 5 years of bleached kraft mill effluent discharge and discharge divertion to sea 94	8) M. Rissato, D. Botelho, A. Rodrigues, E. Foresti, E. Cleto (Brazil). Sulfidogenic anaerobic treatment to remove organic matter, sulfate and AOX from Kraft pulp mill wastewater 26
13:00 - 14:00	Lunch	

TUESDAY, JANUARY 10		
	Sessions Fate and Effects Investigation of Causes	**Sessions IWA Operation of treatment processes**
14:00 - 14:30	4) R. Rutherford, P. Scott, C. Milestone, A. Lister, M. Hewitt, D. MacLatchy (Canada). Androgens and fifth effect evaporator condensates do not cause similar responses in mummichog (*Fundulus heteroclitus*). 95	9) M. Mokua, O. Orori, M. Mwamburi, K. Balozi, K. Senelwa, L. Etiégni (Kenya). Electrocoagulation of a paperboard mill wastewater using varying levels of tin oxide coating on aluminum electrodes. 27
		Sessions IWA Resource and energy recovery or conservation
14:30 - 15:00	5) K. Munkittrick, M. McMaster, L. Bowron, G. Tetreault, M. Hewitt (Canada). Investigation of impacts on white sucker (*Catostomus commersoni*) populations from Jackfish Bay to over 20 years of changes in pulp mill effluent discharges. 96	1) M. Vila, T. Mato, C. Kennes, M. Veiga (Spain). Polyhydroxyalkanoates production from wood mill effluents. 28
15:00 - 15:30	6) P. Martel, T. Kovacs, B. O'Connor, M. Hewitt, D. MacLatch, M.McMaster, J.Parrott, M. van den Heuvel, G. Van Der Kraak (Canada). Progress status of a collaborative research project on the causes and solutions for effects on fish reproduction in Canada 97	2) M. Badshah, W. Parawira, B. Mattiasson (Sweden). Effect of hydrolysing paper and pulp fibre sludge with enzymes on biogas production in anaerobic batch reactors. 29
	Sessions Fate and Effects Monitoring Programs	
15:30 - 16:00	3) F. Niell, R. Carmona, M. Ruiz (Spain). The use of easy measurable variables in the sediment to assess impact in marine sediments affected by organic wastes (1978-2010) 98	3) M. Sandberg, A. Rixen (Sweden). Chemical flocculation as pretreatment for energy-efficient biological treatment. 30
16:00 - 16:30		4) T. Ericsson, Å. Sivard, H. Norrström (Sweden). Process integration of effluent treatment techniques – a way of decreasing the energy and resource use in pulp and paper industries. 31
16:30 - 17:00	Coffee Break	
17:00 - 18:00	Plenary 10 IWA and 8 Fate. Karl Lorber, Institute for Sustainable Waste Management and Technology (IAE) University of Leoben, Austria. Resourcemanagement in the forest industry. 6	
20:00	Symposium dinner	

WEDNESDAY, JANUARY 11		
	Sessions Fate and Effects Monitoring Programs	**Sessions IWA Fate and effect of specific compounds during wastewater treatment**
08:30 - 09:00	1) J. Tana, F. Teixeira de Mello, I. Gonzalez, N. Vidal, A. D'Anatro, D. Larrea (Finland-Uruguay). Monitoring of fish community in the recipient of a greenfield pulp mill at River Uruguay during 2007-2010. 99	1) M. N. Cabrera, S. Toivakainen, M. Martikka, O. Dahl (Uruguay). Chlorate behaviour in Kraft pulp wastewater treatment systems. 32

WEDNESDAY, JANUARY 11		
09:00 - 09:30	2) M. McMaster, J. Parrott, M. Hewitt, T. Kovacs, D. MacLatchy, P. Martel, B. O'Connor, M. van den Heuvel, G. Van Der Kraak (Canada). The environmental effects monitoring program for the pulp and paper industry in Canada: national patterns resulting in IOC for reduced investment into reproduction. 100	2) C. Lucio da Silva, A. Mounteer, T.Duarte, S. Pereira, G. Mozer (Brazil). Organic load removal mechanisms in a eucalyptus bleached kraft pulp mill activated sludge plant. 33
09:30 - 10:00	3) R. Barra, G. Chiang (Chile). The monitoring requirements for assessing river impacts of pulp mill effluents in Chile: some recommendations. 101	3) G. Gielen, K. McGrouther, D. Gapes, T. Stuthridge (New Zealand). Towards understanding colour behaviour in pulp and paper mill wastewater treatment systems. 34
10:00 - 10:30	Coffee Break	
10:00 - 17:30	Poster Display	
	Sessions Fate and Effects Monitoring Programs	**Sessions IWA Fate and effect of specific compounds during wastewater treatment**
10:30 - 11:00	4) M. Salamanca, C. Chandia, A. Camaño (Chile). COP´s spacetemporal distribution in sea water, marine sediments and marine organisms in the coastal platform off Itata river discharge area, Octava Region, Chile. 102	4) A. Klose, T. Hoffmann, E. Tiedtke (Germany). Influence of nonionic surfactants and biocides on the anaerobic stages of paper mill waste water treatment plants. 35
11:00 - 11:30	5) E. Hernández-Miranda, Á. Araneda, E. Díaz, R. Veas, K. Ibáñez, L. Lozano, R. Quiñones (Chile). Biodiversity of the coastal zone off the Itata river mouth after 5 years of effluents release from a pulp mill. 103	5) A. Cunha Lopes, A. Mounteer, T. Valverde (Brazil). Biological activity of bleached kraft pulp mill effluents before and after activated sludge and ozone treatments. 36
11:30 - 12:00	6) M. Dabezies, G. Ferrari, J. Clemente, L. Boccardi, C. Saizar (Uruguay). Biological baseline and monitoring of pulp mill activities at Uruguay river. 104	6) J. Fernandez, N. Diaz (Chile). Abatement of 4-chloropenol under visible light with peroxymonosulfate (PMS) in presence of Co2+. 37
12:00 - 13:00	Lunch	
	Sessions Fate and Effects River / Marine environment impacts	**Sessions IWA Management of waste solids and their evaluation as by-products**
13:00 - 13:30	1) R. Carmona, F. Niell, M. Ruiz, R. Sánchez de Pedro (Spain). A reassessment of Fucacean communities next to a cancelled waste in the Northwest of Spain. Effects of the improvement of the environmental protection technology. 105	1) A. Rezende, M. Diocaretz, J. Castañeda, M. Grandon, G. Vidal (Chile). Waste management: a review of the state of art of pulp and paper mills in South 38 America.
13:30 - 14:00	2) M. Gutiérrez, S. Pantoja, P. Ampuero, E. Tejos, R. Quiñones (Chile). Degradation capability of biopolymers in the coastal environment off central Chile (36.5 S). 106	2) V. Rodrigues, C. Mudado, A.Carneiro, A. Rezende, W. Quirino, L. Santos, G. Ikawa (Brazil). Co-firing of biological sludge and wooden residues from pulp mills for energy production. 39

WEDNESDAY, JANUARY 11		
14:00 - 14:30	3) L. Saavedra, R. Quiñones, J. Becerra (Chile). Phytosterols distribution in coastal and river sediments in central-southern Chile. 107	3) S. Guptaa, S. Chakrabarti, S. Singha (India). Minimisation of hazardous biosludge from pulp and paper industry by ozone oxidation. 40
14:30 - 15:00	4) F. Niell, C. Casas, R. Carmona, M. Ruiz (Spain). More than fourty years of coevolution in a multiple 108 environmental system. Factories, cities and estuarine ecosystems. Is it possible a steady-state in the whole?	4) L. de Caux, L. Dalvi, M. Fonseca (Brazil). Investigation of thermal, physical and chemical characteristics of solid waste from cellulose pulp industry. 41
15:00 - 15:30	5) M. Karlsson, J. Malmaeus (Sweden). Declining levels of dioxins in sediment and fish in the receiving waters of two Swedish bleached kraft pulp mills. 109	5) A. Namazi, G. Allen, C. Jia (Canada). Microwave-assisted chemical activation of pulp mill sludge with potassium hydroxide for endocrine disrupting compounds removal. 42
15:30 - 16:00		6) F. Gallardo, M. Flores, M. Diez (Chile). Effects of the successive application of stabilized secondary sludge from the cellulose wastewater treatment plant to degraded chilean volcanic soils. 43
16:00 - 16:30	Coffee Break	
16:30 - 17:00	Plenary IWA. Claudio Zaror, Chemical Engineering Department, Engineering Faculty, University of Concepción, Chile. The forestry industry and the water issue. 7	
17:00 - 17:30	Plenary Fate and Effects. Renato Quiñones, Faculty of Natural and Oceanographic Sciences, University of Concepción, Chile. Pulp mill, stakeholders' participation and ecosystem research: the case of the PIMEX program. 8	

THURSDAY, JANUARY 12	
09:00	Technical visit

10th IWA Symposium on Forest Industry Wastewaters

POSTERS (Auditorio Emprendo)

Life Cycle Assessment and ecological footprints (water and carbon footprint)

Internal control and system closure

Operation of treatment processes to minimize discharge and nutrient

Resource and energy recovery or conservation

Fate and effect of specific compounds during wastewater treatment

Management of waste solids and their evaluation as by-products

8th International Conference on the Fate and Effects of Pulp and Paper Mill Effluents

POSTERS (Auditorio Emprendo)

Ecotoxicology of effluents and the receiving environment

1.- M. Saavedra, G. Chiang and R. Barra (Chile). Point sources of impact on the Biobío River, Chile: Pulp and paper mill effluents and sewage treatment plant effluents, SIMILAR MAGNITUDE RESPONSES? 113

2.- E. Bay-Schmith, J. Silva, A. Cifuentes, C. Suárez and R. Urrutia (Chile). Ecotoxicity assessment of the industrial effluent from the Valdivia cellulose pulp mill and of its discharge receptor body (río Cruces). 114

Fate modeling / Progress in bioactive chemicals characterization

3.- J. Malm, J. Mikael and M. Karlsson (Sweden). The development of a plume dispersion model for process water discharges. 115

Invertebrate / Fish responses

4- A. Araneda, G. Chiang, A. Valenzuela and R. Barra (Chile). Qualitative characteristics of blood cells and evaluation of hematological parameters in Trichomycterus areolatus (Valenciennes, 1846) and *Percilia gillissi* (Girad, 1854) exposed to pulp mill effluent in the Itata River (Chile). 116

5.- M. Díaz, C. Verdugo, K. Ovalle, F. Correa and R. Figueroa (Chile). Pulp effluent effect on benthic macroinvertebrates community of central Chile. 117

Monitoring Programs

6.- O. Parra, R. Figueroa, C. Valdovinos, E. Habit, H. Cid and C. Zaror (Chile). The environmental significance of forest industry effluents (Pulp and Papers mills) in the Biobío river basin. 118

7.- P. Torres, R. Jerez, S. Figueroa and D. Arcos (Chile). Short term coastal environmental monitoring of a local pulp mill plant treatment outfall. 119

Nutrient enrichment

Proteomics, Genomics, Metabolomics

Reproductive effects / Endocrine disruption

8th International Conference on Fate and Effects of Pulp and Paper Mill Effluents

IWA International Water Association

10th IWA Symposium on Forest Industry Wastewaters

JOINT PLENNARY

JOINT PLENARY

Where are the opportunities for our environmental technologies as the forest processing industry moves towards a biorefinery

D. Grant Allen[1], Daniel Gapes[2], Michael Jack[2], Robert Lei[2] and Trevor Stuthridge[2]

[1]*Department of Chemical Engineering and Applied Chemistry*
University of Toronto, Toronto, ON email: dgrant.allen@utoronto.ca
[2]*SCION, Rotorua, New Zealand*

The world's interest in sustainable processing and global warming provides great opportunities and challenges for the forest product industry. As a result it is likely that we will continue to see our pulp and paper mills and wood products facilities transform themselves into biorefineries that can produce a wider range of products (energy, materials, etc.). There is little doubt that the processing of residual streams within such biorefineries will be central to their economic and environmental performance. Within this context, we expect that there will be considerable need to develop and enhance technologies associated with the conversion of waste streams into valuable products and mitigate their impact on the environment. These so called 'environmental technologies' include everything from advanced separation processes through to traditional wastewater treatment processes. In this paper we will present the findings of a literature review exploring what are some of the likely key environmental technologies in current and future biorefineries and what are some of the challenges in their implementation. We will also consider what might be the role of the IWA Specialist Group on Forest Industry Wastewaters in this emerging area.

Keywords: Biorefinery, Environmental Technology, Anaerobic Treatment, Separation Technologies, Industrial Water Recycling, Recovering Value

JOINT PLENARY

Investigating the causes of mill effluent effects on fish reproduction in Canada: lessons learned

M. McMaster

Aquatic Ecosystem Protection Research Division Environment Canada, Burlington, ON Canada

The national assessment of the first three cycles of the Canadian Environmental Effects Monitoring (EEM) Program showed that pulp/paper mill effluents were causing general nutrient enrichment in the receiving environment as well as metabolic disruption in fish. The consequence of metabolic disruption is that fish are allocating less energy towards reproduction. This is evidenced by larger condition factor and liver size as well as smaller gonads in fish inhabiting effluent-containing waters. When effects are identified, the EEM program calls for Investigation of Cause (IOC) followed possibly by Investigation of Solution (IOS) studies. In the case of metabolic disruption, the causative agent(s) were not known and intensive IOC and IOS studies were required. Various IOC and IOS studies have been conducted on metabolic disruption by different institutions over the past decade, with mill specific progress. Since 2005 a multi-agency consortium involving industry, government and academic researchers was formed around a centralized project. As the EEM Program is now being applied to metal mines and possibly municipal sewage plants in Canada, this project is serving as a model for these sectors. It may also serve as an example to South American countries (e.g. Chile and Brazil) that have expressed an interest in the EEM Program and are looking to incorporate effects-based monitoring into regulatory frameworks. A retrospective look at these investigations, including factors contributing to the most recent successes and lessons learned, will be presented.

Keywords: pulp mill effluent, fish reproduction, investigation of cause, investigation of solution, endocrine disruption

JOINT PLENARY

Monitoring of the impacts of pulp mill effluents – a historical overview, challenges and lessons learned

Jukka Tana

ÅF-Consult Ltd, Rajatorpantie 8, FI-01600 Vantaa, Finland Ph + 358 4 3485644 email jukka.tana@afconsult.com

The impacts of pulp mill effluents have been monitored almost as long as the mills have been operating. The objectives of monitoring could be summarized in four different points: i) to know; ii) to be able to follow the possible impacts of the effluents; iii) to be able to mitigate and iv) to be able to inform. The first monitoring programs consisted of few effluent samples together with randomly collected water samples from the receiving water. The development of national and international environmental legislations and the general environmental awareness increased the pressure of information of industrial effluent effects. As a consequence the development of environmental monitoring programs and the overall scientific interest for different impacts of industrial effluents increased. In this respect the effluents from pulp and paper mills have been by far the most studied. Different countries have had different strategies for the effluent monitoring, but all have the same objective to be able to improve the quality of the effluents and to decrease the impacts.

The aim of this presentation and paper is to give a short overview in the development of the environmental effect monitoring of pulp and paper mill effluents and to compare monitoring programs between different countries with significant pulp and paper production. The relationship and interactions between the technical development and the results of monitoring studies is also discussed. Also this paper tries to give information on the lessons learned during the development of monitoring activities and how the results of the scientific research have been implemented in the effect monitoring.

Keywords: monitoring, pulp mill effluent, water quality, environment, effects

JOINT PLENARY

Resource-management in the forest industry

Karl E. Lorber and Renato Sarc

Institute for Sustainable Waste Management (IAE), Montanuniversitaet Leoben, Austria email: iae@unileoben.ac.at

In Austria, more than 47% of the total surface area is covered by forest, making forest industry to an important industrial sector. The utilization of biomass as well as energy recovery from wood residues have developed to an essential contribution to the national resource management. Apart from construction and furniture industry as well as pulp and paper production, the renewable resource wood plays an important role as biofuel in the national energy scenario. In rural areas, the bio fuels, wood chips and wood pellets are increasingly displacing oil and gas heating in households, and waste wood and wood residues (like bark and saw dust) are increasingly recovered as waste fuel or quality controlled solid recovered fuel (SRF) to be used in co-incineration plants. As an example, technical features as well as practical experiences of a modern 30 MW-biomass power plant are reported, where residues like bark and saw dust are utilized for co-generation of heat and power, by applying the so-called Organic Rankine-Cycle (ORC) process. Finally, utilization and disposal of combustion residues (e.g. fine and coarse ashes) are reported and new developments in upgrading technologies for biofuel (e.g. gasification and conversion to syn-gas, Fischer-Tropsch synthesis, etc.) are discussed.

Keywords: Biomass, Biofuel, Wood Residues, Wood Pellets Heating, Solid Recovered Fuel (SRF), Organic Rankine-Cycle (ORC), Ash Utilization and Disposal

JOINT PLENARY

The forestry industry and the water issue

Claudio Zaror

Chemical Engineering Department, Engineering Faculty, University of Concepción. P.O. Box 160- C. Concepción-Chile. E-mail: czarorl@udec.cl.

The forestry industry worldwide undergoes increasing pressures to reduce environmental impacts and, at the same time, contribute to sustainable development. In particular, effects on water resources appear as a key issue in areas impacted by climate change, where reductions in rainfall and increases in temperature pose a serious risk to industrial operations.

Forestry activities may conflict with local communities due to the effect of plantations on scarce local water sources. In this respect, the effect of forestry plantations on water balance around a river basin should be fully understood, in order to design appropriate forestry management strategies and prevent any negative impact on water bodies.

Additionally, cellulose production involves water demanding operations and water availability may become a critical limiting factor for further development. Water consumption may need to be reduced below current standards in areas with increasing water stress. Although the specific water consumption has been drastically reduced in the last two decades, the scale of production per plant has almost trebled, leading to high fresh water uptakes. Moreover, stricter environmental legislation has targeted new trace contaminants that need to be controlled before discharge to water bodies.

A number of technological challenges associated with circuit closure and advanced effluent treatment remain unsolved. Application of new technologies, such as membrane separation processes aimed at water and chemical resources recovery, is still a pending issue. Although these technologies have been proven in other industrial sectors, applications in cellulose mills are limited.

Finally, the economic consequences of delaying the implementation of more aggressive strategies on water management in the forestry industry are highlighted here.

Keywords: water issue, forestry industry, circuit closure, climate change

JOINT PLENARY

Pulp mill, stakeholders' participation and ecosystem research: the case of the PIMEX program

Renato Quiñones

Programa de Investigación Marina de Excelencia (PIMEX), Faculty of Natural and Oceanographic Sciences, University of Concepción, P.O.Box 160-C, Concepción, Chile. E-mail: rquinone@udec.cl

The construction of major industrial projects can generate important controversy among stakeholders and sometimes significant social conflict. Recently, the advantages of including the full range of stakeholders in dealing with environmental problems in a manner that is informative, iterative, and interactive has become evident. Frequently, there is very limited scientific knowledge of the ecosystems where industrial operations are planned. Consequently, debates and negotiations among stakeholders are strongly driven by perceptions and values. When there is substantial knowledge of the ecosystem, science is able to significantly reduce the uncertainty related to environmental impacts. However, science is never able to totally eliminate this uncertainty, given the immense complexity of socio-ecological systems. Here, I analyze the role of science in the interaction among stakeholders of the marine ecosystem of the Nueva Aldea pulp mill plant (ARAUCO; central-south Chile). This plant has a 50.8 kilometers pipeline that carries fully-treated effluents to the coastal zone and has an underwater extension of 2.3 kilometers. To ensure that environmental concerns regarding the pipeline are adequately addressed, the Regional Government, ARAUCO, the Federation of Artisanal fishers, local artisanal fishers unions, and the University of Concepción put together a multi-stakeholder forum called the "Nueva Aldea Fishery round table" (NAFRT)in December 2005. This forum has been working continuously to enable the direct and positive dialog among stakeholders, with the goal of reaching a consensus regarding the operation and management of the pipeline. In addition to the environmental monitoring program required by Chilean environmental legislation, the University of Concepción and ARAUCO agreed to conduct a major research program on the coastal marine ecosystem. The research program design was based on public concerns expressed during the Environmental Impact Assessment process, and on scientific advice regarding key research areas for understanding ecosystem structure and functioning. Thus, in winter 2006 the program known as PIMEX-Nueva Aldea (Program of Marine Research of Excellence on the Ecosystem surrounding the Mouth of the Itata River) was born. This research program has also been approved by stakeholders participating in the NAFRT. PIMEX has a Directive Board composed by representatives of all NAFRT stakeholders. A general assessment of the achievements, difficulties, scientific contributions, and lessons learned by PIMEX is presented.

Keywords: ecosystem research, sustainability, stakeholder interaction, pulp mill effluent, the role of science

International Water Association

10th IWA Symposium on Forest Industry Wastewaters

ORAL PRESENTATIONS

LIFE CYCLE ASSESSMENT AND ECOLOGICAL FOOTPRINTS (WATER AND CARBON FOOTPRINT)

What do water profiles, Water Footprints, and water stewardship assessments reveal about the environmental sustainability of the forest products industry?

P. Wiegand[1], C. Flinders[1], G. Ice[1], D. Sleep[1], B. Malmberg[1], I. Lama[1], J. Schwartz[2], M. Mensink[3], D. Phenicie[4] and W. Larson[5]

[1]*National Council for Air and Stream Improvement, Inc. (NCASI);* [2]*American Forest and Paper Association (AF&PA);* [3]*Confederation of European Paper Industries (CEPI);* [4]*Council of Great Lakes Industries;* [5]*LimnoTech*

Since the 2009 IWA conference in Fredericton, New Brunswick, Canada, there has been a rapidly accelerating exchange of ideas about environmental sustainability, and water has been featured prominently in that exchange. At last count, there were approximately 20 different global initiatives focused primarily on exploring and/or promoting sustainable management of fresh water resources. Because of its reliance on fresh water, the forest products industry has been increasing involved with these efforts. NCASI has prepared separate water profiles for the North American, Canadian, and European forest products industries. The industry has partnered with several organizations to define the metrics relevant to water footprinting and stewardship at businesses, and has undertaken various assessments, including pilot evaluations, of the protocols used or under development by several leading Environmental Non-Governmental Organizations (ENGO).

This presentation will summarize important findings from these interactions and investigations including: (a) the influence of forest management on water resource availability and quality, (b) amounts of water sourced to, discharged from, and consumed by forest products manufacturing operations, (c) what water footprints and stewardship assessments reveal (and do not reveal) about water influenced by the forest products industry, and (d) information likely to be required by companies to support a comprehensive water sustainability assessment. The presenter will also comment on future activities and the role the forest products industry may play in fostering a greater understanding of sustainable water resource management.

Keywords: consumption, footprint, forest products, profile, stewardship, sustainability, water resources

Effects of further effluent treatment stages at pulp and paper industries

Å. Sivard[1], M. Malmaeus[2], M. Almemark[2], M. Karlsson[2], T. Ericsson[1] and O. Simon[1]

[1]ÅF, Frösundaleden 2, S-169 99 Stockholm, Sweden Ph +46 10 505 1307, fax +46 10 505 0010, E-mail asa.sivard@afconsult.com, tomas.j.ericsson@afconsult.com; olle.simon@afconsult.com. [2]IVL Swedish Environmental research Institute, Box 210 60, S-100 31 Stockholm, Sweden
E-mail:mikael.malmaeus@ivl.se, mats.almemark@ivl.se, magnus.karlsson@ivl.se

The consequences of upgrading existing secondary treatment system with an additional treatment stage at two Swedish pulp and paper mills have been evaluated. Calculations of discharge reductions and costs for membrane filtration, chemical precipitation and sand filter as tertiary treatment have been performed. Nutrient discharges are reduced best by chemical precipitation and suspended solids are reduced most efficiently by filtration. Investment and operating cost are lowest for sand filters.

The recipient of one of the mills is a big lake in the middle of Sweden and the recipient of the other mill is the Gulf of Bothnia. Simulations did not show any significant environmental improvement in the receiving waters due to the reduced discharges.

Analyses of the total environmental impact (LCA) of the new treatment stages show that the total nutrient discharge is reduced in all cases. However, the total environmental impact in all other aspects increased due to increased energy consumption, increased chemicals consumption at operation and production and increased sludge production in the new treatment stages. There is no standard way of comparing these categories, but different ways of evaluation and normalization compared with total emissions per year and person in a certain region can be used or comparison with political goals for different categories.

The results show that it is not at all obvious that the added treatment stages will give a better environmental status based on all aspects.

Keywords: membrane filtration, chemical precipitation, sand filters, pulp & paper mill effluent, LCA, environmental impact

STATUS OF ENVIRONMENTAL ISSUES NATIONALLY AND REGIONALLY

An update on the development of numeric nutrient criteria in the United States: a forest industry perspective

D. McLaughlin
National Council for Air and Stream Improvement, Inc. Northern Regional Ctr., Kalamazoo, MI, USA.
E-mail: douglas.mclaughlin@wmich.edu

Considerable effort has been spent by state and federal environmental agencies and nongovernmental organizations in the United States in recent years related to develop numeric water quality criteria for nitrogen and phosphorus. Driven by concerns over the negative water quality impacts that can result from excessive levels of these nutrients, the development of numeric criteria has had a prominent place in debates over the most appropriate and scientifically defensible approaches used to prevent and manage nutrient pollution.

Recently, numeric nutrient criteria have been adopted for inland fresh waters in two states that are significant to the forest products industry, Wisconsin and Florida, and efforts to develop and finalize numeric nutrient criteria in other such states continue. One of the most significant challenges in developing numeric nutrient criteria that are useful in water quality management is the high degree of variability frequently observed between nutrient concentrations and associated measures of biological or ecosystem response such as algal growth or changes in macroinvertebrate and fish communities. Relatively simple approaches for criteria development may result in criteria that are poor predictors of biological or ecological condition, leading to management decisions that have a high probability of error.

This paper describes several approaches being taken in a number of forest product industry states in the US to manage nutrient pollution. It includes information on the types of numeric nutrient criteria that have been adopted or are being considered. The use of decision error analysis to evaluate candidate numeric nutrient criteria is also described.

Keywords: water quality criteria, regulation, nutrients, forest products industry, decision error

Comparative analysis of wastewater technology and river environmental impact: Kraft pulp mills in South America (Chile and Uruguay)

S. Videla and C. Muñoz

AMEC, Environment & Infrastructure , Santiago-Chile. E-mail: sebastian.videla@amec.com.

Two large Kraft pulp mills located in Chile and Uruguay generated social, environmental and political conflicts due to the site in which they were located, near water resources and ecological areas. These two cases are described in this work, including the most relevant environmental issues and the wastewater technologies used to reduce the environmental impact. The main objectives of this review are the comparative analysis of two different concepts of technology used in these mills. The wastewater treatment systems were designed specifically for basin characteristics of each area. The Chilean pulp mill initially included primary, secondary and tertiary wastewater treatment units. The last unit was designed to reduce colour in the effluent using aluminum sulphate. From the start up, the discharge was sent to a wetland with a rich ecosystem. At the same time it was observed that the black neck swan population had been reduced causing a controversy regarding the influence of the mill. After this conflict, a quaternary process was added to the wastewater system in order to improve the final quality of the discharge. In Uruguay, the mill was designed with primary and secondary treatments; the latter corresponding to a conventional activated sludge process. The discharge was in the Uruguay River that is the border between Argentina and Uruguay. The main conflict was generated from the Argentinean people living near in front of the mill. Both cases correspond to the most critical environmental and social conflicts occurring in the last twenty years in the forest industry.

Keywords: Kraft mill, Environment, South America

Characterization of effluents from A-EP-DND and A-EP-D-P eucalyptus pulp bleaching sequences

M. Cabrera[1], L. Clavijo[1], M. Martikka[2] and O. Dahl[2]

[1]Forest Process Engineering Group- Engineering School- Universidad de la República- Uruguay. Email: ncabrera@fing.edu.uy. [2]Clean Technologies Group- Forest Products Technology Department- Aalto University-Finland

In this work the analysis of the filtrates obtained from the optimization of the conditions of novels eucalyptus pulp bleaching sequences A-EP-DND and A-EP-D-P were done. In this regard were measured BOD7, COD, TOC, Chlorate and AOX among others parameters. These sequences increase the possibilities of internal water circulations within the bleaching plants and hence permit to reduce the amount of water used.

Uruguayan eucalyptus industrial pulp after oxygen delignification was used. First, the sequence was optimized, being the principal optimization parameters the Acid stage's reaction time and ClO2 charge in D stages. The optimized sequence was compared against A/D-EP-D-P used as reference sequence. The novel sequence use less chlorine dioxide, thus, half of the chlorate formed in the sequence used as reference is formed in the new one. AOX formation in the new sequence is just 33 g/ton, lower than the 48 g/ton formed in the reference sequence and lower than the AOX formation reported in literature for different eucalyptus bleaching sequences. COD; BOD7 and TOC levels are in the same order as the reference sequence, however 70 - 75 % of the total effluent load in TOC and COD corresponded to the pre-bleaching filtrates. Considering that the only stage where chlorine compounds are used is the D1, the filtrates from the A stage (unlike the A/D stage) could be reused within the mill and therefore the major part of the organic load to the wastewater plant is deviated. Consequently, this new bleaching sequence is very promising.

Keywords: Eucalyptus pulp, novel bleaching sequence, system closure, AOX formation, Chlorate formation

Effects of xylanase application on Eucalyptus Kraft-O Pulp

M. Borges[1], C. Mudado[2], J. Colodette[2] and R. Chaves de Oliveira[2]

[1]Universidade Federal de Viçosa / Fibria MS Celulose Sul Mato-grossense. [2]Universidade Federal de Viçosa

Currently, the use of enzymes appears to be a promising approach for clean bleaching processes and the reduction of bleaching chemical consumption. However, an increase in the organic matter content of effluents has been observed when enzyme stages are used. In this work, the quality of the filtrates and the pulp were characterized after the enzymatic pre-bleaching stage. Enzymatic pre-bleaching of eucalyptus pulp was performed under four different conditions using xylanase and a reference. Following each pre-bleaching stage, the properties of the pulp (kappa number, brightness, hexenuronic acid, xylan removal and pulp yield lost) and of the resulting effluents (COD, BOD, color and conductivity) were determined. In addition, a complete bleaching sequence was performed to characterize the pulp quality at the end of the sequence. Finally, the pulp was refined in a PFI mill, and the mechanical and physical properties of the resulting paper were determined. The best pulp quality was achieved at pH 7, when smaller kappa number, higher pulp brightness and smaller amounts of hexenuronic acid were observed. However, at this condition a bigger yield loss and a higher effluent COD load were detected.
The bio-treatment of all effluents presented high COD removal (> 85%). After the complete bleaching sequence the pulps treated with enzymes presented higher brightness and lower kappa number; these changes were modest, though. The enzymatic dosage of 200 g adt-1 at pH 7 presented the highest viscosity increase. The enzymatic treatments presented small changes in the mechanical and physical properties of the produced paper handsheets.

Decolorization of black liquor by 9 strains of lignocellulolytic basidiomycetes

V. Da Re and L. Papinutti

Laboratorio de Micología Experimental (PROPLAME-PRHIDEB), Departamento de Biodiversidad y Biología Experimental, PabII Ciudad Universitaria, (C1428EHA) Ciudad de Buenos Aires, Argentina.

Many pulp mills discharge into the environment their pulping effluents with insufficient or no treatment causing a dramatic increase in environmental pollution particularly to aquatic ecosystems. Black liquor from Kraft pulping process contains a blend of alkali lignins, its intense brown color along with its recalcitrance to biodegradation make this residue an issue of environmental concern. Dark brown color is not only aesthetically unacceptable but could also inhibit the process of photosynthesis in natural aquatic environments due to the barrier effect to sunlight. Bioremediation strategies may be useful to reduce the intensity of color (1), however, the high concentration of salts is a strong inhibitor for microorganism growth or for their enzyme activities. The objectives of this study was to select fungal strains capable of growing in the harsh conditions inflicted by the presence of black liquor and to be able to decolorize it. The nine fungal strains studied were 7 belonging to the white-rot group: *Coriolus antarcticus*, *Peniophora* sp., *Pycnoporus sanguineus*, *Trametes elegans*, *T. trogii*, *T. villosa*, *Steccherinum* sp., and one strain belonging to the brown-rot group: *Laetiporus sulphureus*.

Previous to use the pH of black liquor (originally was 13) was adjusted to 6 with citric acid 1M. Agar plates containing malt extract medium supplemented with 4% (vol/vol) black liquor were inoculated with the nine strains. Plates were incubated at 28°C and monitored each 24 h for growth and decolorization. Growth was followed by measuring radial extension of mycelium. Ligninolytic enzymes from samples of decolorized and non-decolorized zones were measured.

A decolorized halo appeared when the fungus degraded the black liquor. Intensity of decolorization in 30-day-old plates was estimated by measuring luminance parameter. Luminance (L*) was measured against a white reference standard (titanium oxide (L* = 94.5)) with a ColorMunki Portable Spectrophotometer (D65). The color of the samples was described according to the CIE L*a*b* color system, where L*, a* and b* are the coordinates of the color in the cylindrical color space, based on the theory that color is perceived by black-white (L* = lightness-darkness), red-green (a*), and yellow-blue (b*) sensations.
Cultures exhibited a variable extent of black liquor decolorization depending on the culture day and fungal strain. Four of the nine strains assayed showed to be able to significantly decolorize the black liquor: *L. sulphureus*, *P. sanguineus*, *T. elegans*, *C. antarcticus*. The highest decolorization value (L*=52.77), along with the highest velocity, was achieved by *L. sulphureus*. Interestingly, this fungus was described as brown rot causing and unable to produce ligninolytic enzymes. However, several findings provided information about attack of lignocellulose by brown rot fungi through production of hydroxyl radicals, which is a strong oxidizing agent. Possibly, this mechanism used by the fungus to degrade cellulose and to modify lignin in nature should be responsible of the decolorization observed in plates. On the other hand, one lignin-modifying enzyme (MnP) was detected in all white-rot fungi with a significative decolorization value.

Keywords: black liquor, decolorization, laccase, CIE L*a*b*, fungi

Water and wastewater management in modern bleached Eucalyptus Kraft pulp mills

O. Pikka and M. Grötzner

Andritz Pulp and Paper – Brazil, Andritz Brasil Ltda – Curitiba-Paraná. E-mail: mariana.grotzner@andritz.com

Pulp and paper industry is a big water consumer and effluent generator. The availability of fresh raw water dictates the location of production unit. Water consumption and effluent generation are very relevant issues for bleached eucalyptus Kraft pulp production. The scarcity of usable water, environmental restrictions for effluent disposal and the importance of pulp production in the economies justify the studies in the area. Moreover, the water and wastewater are of great interest, since water footprint is coming under especial attention. Improvements have been reached over the years for the water specific consumption, see Picture 1. In general, modern pulp mills may consume 15-20 m^3 of fresh water per ton of pulp. The reduction of water usage and effluent amount could be obtained through system water loop closure and process internal recirculation. Closure of water circuit has contributed for the development of water and wastewater management actual concept. This theme has showed that it is not possible to close completely the water cycle due to the undesirable components accumulation in fiberline and recovery processes. Effluent treatment and purification are removing or reducing undesirable compounds from the wastewater. Very recent research has shown that purified effluent could be used as process water in fiberline system, see Picture 2. According to laboratory tests, purified effluent can be re-circulated to the process without impact on pulp quality, chemical consumption and significant accumulation of substances. The new concept of reutilization of treated effluent looks very promising step in reduction of fresh water consumption in pulp mill. This paper presents not only the evaluation of state-of-art of water and wastewater management in modern pulp mill, but mainly, new ideas of reducing, recirculating and reusing process water.

Keywords: water, wastewater, reduce, re-circulate, reuse, Kraft Pulp Mill

OPERATION OF TREATMENT PROCESSES TO MINIMIZE DISCHARGE AND NUTRIENT

Use granular aerobic sludge MBR (membrane bioreactor) to treat paper mill effluent

I. Horta[1], C. Mudado[1], C. Piacsek[2] and V Lopes[1]

[1]*Departamento Eng. Florestal –Federal University of Vicosa, Brasil. E-mail: mudado@ufv.br.*
[2]*Departamento de Eng. Quimica – Universidade Federal do Rio de Janeiro*

The activated sludge process is widely used for the treatment of industrial effluents, achieving high rates of organic matter removal. Nevertheless, the separation of biological sludge in secondary clarifiers can be a serious problem of this process due to the poor floc formation, causing solids losses and, consequently, deterioration of treated effluent quality. Membrane bioreactors (MBR) emerged as an alternative to solve this problem, since the separation of suspended solids from the treated effluent is not dependent on the sludge settling characteristics. However, it has been observed that malformed flocs foul the pores of the membranes, causing the reduction of flux and consequently an increase in the need for membrane cleaning. Thus, even in MBR, good formation of the biological floc is essential for proper operation of the process. Recently, researchers working with aerated activated sludge reactors found a different organization of the microbiota in the form of aerobic granules. In this configuration, the bacteria clump together more tightly, thus reducing the possibility of biofilm formation on the membrane surface, and allowing the maintenance of a higher flow for longer periods of time during filtration. This study examined the possibility of formation of aerobic granules with paper machine effluent and compared the removal efficiency of COD and BOD of the granular sludge with the conventional flocculent sludge. The filterability of sludge granular and flocculent was also compared. Two reactors were operated in parallel, one with aerobic granular sludge and the other with flocculent sludge. Both systems achieved BOD removal efficiencies of 97% and COD removal of 89% and 91% for granular and flocculent sludge, respectively. In the granular sludge reactor, granules of different sizes with diameters up to about five millimeters, visible to the naked eye were formed. Sludge filtration tests showed a higher critical flux for granular sludge with total suspend solids (TSS) equal to 2700 mg l^{-1}. However, the opposite behavior was observed when the reactors were operated with a TSS of 4400 mg.l^{-1}.

Keywords: aerobic granule, paper mill effluent, membrane bioreactor

Filamentous bacteria identification and control strategies

J. Liu[1] and M. Paice[2]

[1] *GENIVAR Consultants LP, 500 – 600 Cochrane Drive, Markham, ON L3R 5K3, Canada.*
[2] *FPInnovations, Paprican Division, 3800 Wesbrook Mall, Vancouver, BC V6S 2L9, Canada*

Many activated sludge wastewater treatment plants (WWTPs) in the pulp and paper industry suffer from foaming and bulking problems caused by excessive growth of filamentous bacteria, which is the most common operational disorder in the wastewater treatment industry worldwide. This paper will attempt to correct a common misunderstanding that all filamentous bacteria are the same and therefore the strategy for controlling them is the same. The paper will also illustrate how new identification techniques may help to understand the causes of filamentous proliferation and thereby enable us to establish specific control strategies and identify short- and long-term filamentous bacteria control options.

Microscopic identification keys for filamentous bacteria are available from several sources. However, the majority of the information has been collected from municipal wastewater treatment systems. Industrial effluents tend to be more diverse and give rise to unique predominant filaments depending on individual effluent characteristics. For instance, Cyanophyceae, rarely seen in municipal wastewater treatment systems, can be caused by a deficiency of nutrients and leads to foaming and bulking in some industrial effluents. It is important to identify causative filaments which could relate to operational conditions, and apply this knowledge to foaming and bulking control strategies.

Historically, activated sludge bulking and foaming problems were considered an engineering problem. Therefore, a generic control approach, either chlorination, or polymer addition, or selectors, has generally been utilized to control these problems. Although these short term control measures have successfully reduced bulking and foaming problems in many activated sludge systems, there are also regular reports of failures.

For longer term control, it is critical that species-specific solutions are found for each identified occurrence. Since the strategy is specific to the ecology and physiology of the species, a long term control can be achieved. For example, the most common filamentous bacteria in pulp and paper effluent, type 021N and Thiothrix spp., can be caused by septic conditions in the primary clarifier. Therefore, bringing dissolved oxygen (DO) into the primary clarifier may resolve the problem. Another example is M. parvicela which has been ranked as the number one foam causing filament in the world and second in the USA. Due to its specific physiological properties, anoxic and anearobic selectors will not depress its growth but will, in fact, assit its proliferation.
Despite much research, bulking and foaming sludge continue to be a problem in operating WWTPs. Due to the complexity of the microbial ecology of activated sludge, a single solution control strategy might not be achieved in the very near future. However, correct identification of the bacteria is critical to controlling bulking and foaming problems. This paper will review techniques currently available and strategies to follow once the bacteria are identified.

Control of bulking of sludge in activated sludge process treating pulp and paper mill effluent

S. Chakrabarti, S. Gupta, Methoxy and R. Vardhan

Thapar Centre for Industrial Research & Development, Paper Mill Campus, Yamuna Nagar- 135001, Haryana, India. E mail: sanjeev@thaparpaperresearch.org, Phone: +911732292737, FAX: +911732292748

Bulking sludge and proliferation of filamentous organisms in the effluent treatment process in pulp and paper mills are very common. Generation of filamentous organisms and nonsettlable flocs are closely related to operating conditions. The dissolved oxygen (DO) level affects the morphology as well as performance of the process to a great extent. A bench scale study has been carried out to elucidate the optimum DO concentration in activated sludge process with pulp and paper mill wastewater. Very low level of DO (0.2-0.3 mg/l) promoted the excessive growth of filamentous organisms and the higher organisms like protozoa and rotifers were particularly non-existent; the performance of the process was very poor; reduction in COD and colour was only 25-30% and traces respectively. The performance in the biological treatment improved at DO level of 0.5 mg/l but nature of sludge remained bulking. The moderate level of DO i.e. 0.9 to 1.1 mg/l was found to be adequate for operation of activated sludge process, which is lower than the recommended level of DO (1.5-2.0 mg/l). The morphology of sludge, growth of higher organisms and performance of process remained comparable at higher level of DO i.e. 2.0-2.5 mg/l. The reduction in the COD and colour was 63-65 and 43-47% respectively. The sludge volume index (SVI) at low (0.2-03 mg/l) and moderate (0.9-1.1 mg/l) DO level was >200 and <40 ml/g respectively. The alteration of DO from lower to higher or vice-versa resulted in change of morphology and performance of the process.

Keywords: Activated sludge process, filamentous organisms, flocs, reactor performance, sludge volume index

Effect of hydraulic retention time on the performance and microbial community in a thermophilic aerobic membrane bioreactor

X. Qu[1,2], W. Gao1, A. Chen[2] and B. Liao[1]

[1]*Department of Chemical Engineering, Lakehead University, 955 Oliver Road, Thunder Bay, ON P7B 4H8, Canada.* [2]*Department of Chemistry, Lakehead University, 955 Oliver Road, Thunder Bay, ON P7B 4H8, Canada*

In order to achieve energy reduction and system closure in pulp and paper industry, a new in-mill wastewater treatment technology based on thermophilic aerobic membrane bioreactors (TAMBR) was developed. The TAMBR was operated at 51±1°C for the treatment of high-temperature thermomechanical pulping (TMP) pressate over a 157-day period. The effect of hydraulic retention time (HRT) on the performance and microbial community of the TAMBR was studied. The optimal HRT for the operation of TAMBR was also determined.

The performance of the TAMBR was investigated at three HRTs of 1.7±0.3, 1.1±0.1 and 0.9±0.1d. With the decrease in HRT, the organic loading rate increased from 2.7±0.3 to 3.9±0.3 kgCOD/m^{3}d; the total COD removal efficiency slightly decreased from 92% to 89%, while the supernatant COD increased significantly which may lead to more frequent membrane fouling; and the MLSS concentration also increased from 9.9±0.6g/L to 12.0±0.5 g/L.

The potential change of microbial community during the three HRTs is being studied by polymerase chain reaction (PCR)-denaturing gradient gel electrophoresis (DGGE) technique.

Considering the overall performance of the TAMBR in terms of COD removal efficiency and the membrane fouling as well as the initial reactor construction cost, it is recommended the HRT of 1.1day to be the optimal HRT.

Keywords: Thermomechanical pulping (TMP) pressate, system closure, Thermophilic aerobic membrane bioreactor, Hydraulic retention time (HRT), microbial community

Kinetics of aerobic thermophilic treatment of a kraft pulp bleach plant wastewater

R. Cabrera-Padilla[1], P. Fernandes[2] and E. Pires[1]

[1]Departament of Hydraulics and Sanitation, Engineering School of São Carlos (EESC), São Paulo University (USP), Av. Trabalhador São-carlense 400, São Carlos 13566-590, SP, Brazil. [2]Post-Graduate Program Interunits in Bioengineering, EESC/FMRP/IQSC - São Paulo University (USP), Av. Trabalhador São-carlense 400, São Carlos 13566-590, SP, Brazil.

Recalcitrant compounds found in pulp and paper mill effluents, which are usually generated at high temperatures, have shown the potential for thermophilic treatment. However, the biomass yield in thermophilic processes is usually very low what may prevent the successful application of thermophilic treatment. Using bleached kraft pulp mill effluents; this study aims to evaluate the kinetic parameters of an aerobic sludge submitted to different acclimation strategies. The first reactor, R1, was fed with bleached kraft pulp effluent, while reactor R2 was fed with easily biodegradable synthetic effluent and then the gradual substitution for bleached effluent was performed after the sludge was adapted to the thermophilic environment. Both reactors were operated in batch at 55°C. The second strategy has a slight advantage in COD removal and a considerable higher biomass production. The kinetic parameters (KS, μmax, Y, Kd) for both strategies were determined using the Monod model and their values indicated that the approach used for reactor R2 provided higher reaction velocity.

Keywords: Aerobic; bleached kraft pulp effluent; kinetic constants; sludge; thermophilic

Managing effluent phosphorus by optimizing the activated sludge process in ECF kraft pulp mill

T. Laukkanen, S. Toivakainen, H. Shahid and O. Dahl

Aalto University, School of Science and Technology, Department of Forest Products Technology, Clean Technologies Group, Helsinki, Finland, Tel +358 50 911 8746, sakari.toivakainen@aalto.fi

The mass balance of phosphorus in secondary treatment i.e. the activated sludge process (ASP) is quite simple: the influent mass flow equals that of the effluent and waste sludge (WS) together in a steady state situation. Hence to reduce effluent load one can either decrease the influent load or increase the phosphorus outflow within WS. This presentation concentrates on the latter case in a situation where excess phosphorus for biochemical need in ASP is present. Phosphorus outflow within waste sludge can be increased by increasing the flow of WS itself i.e. reducing sludge retention time (SRT). Reducing SRT on the other hand has its limits in pulp mill ASP because reduction of organic compounds measured by COD or TOC will decrease when SRT is decreased below 15 days. The scarcity of other nutrients like nitrogen can also limit the uptake of phosphorus by hindering biomass synthesis in ASP. Effluent phosphorus load can also be reduced by increasing the phosphorus concentration in WS. This can be done by enhanced biological phosphorus reduction (BPR) or by chemically using simultaneous precipitation. BPR using variations between anaerobic and aerobic phases may be applied especially when phosphorus concentration is elevated for example due to opening phosphorus circulations in the pulp mill. On the other hand the simultaneous precipitation of phosphorus can be carried out quite easily even in an old or heavily loaded ASP by feeding iron salts in to the process and this way immobilize dissolved phosphorus from wastewater into activated sludge. However, the phosphorus reduction achieved with pulp mills' wastewaters has not been as high as that with municipal wastewaters by this method.

Keywords: Activated sludge, phosphorus, sludge retention time, biological phosphorus reduction, simultaneous precipitation , pulp mill effluent

Applicability of biological phosphorus removal process for kraft pulp mill wastewater treatment

H. Shahid, S. Toivakainen, T. Laukkanen and O. Dahl

Department of Forest Products Technology, Clean Technologies Group, Aalto University, PO Box 16300, FI-00076 Aalto, Finland. Tel: +358 505361704, E-mail: shahid.hussain@aalto.fi

Due to the stringent regulatory limits, Finnish pulp mills will require to lower phosphorus concentration in the effluent of wastewater treatment plants. Biological phosphorus removal (BPR) process, a cost-effective alternative for tertiary wastewater treatment, is widely used for the reduction of phosphorus from municipal wastewater. Characteristics of kraft pulp mill wastewaters were found suitable for BPR process. Applicability of BPR process for the treatment of kraft pulp mill wastewaters was studied using four laboratory-scale sequencing batch reactors (SBRs). One SBR was operated according to the conventional fully aerobic activated sludge (AS) principle. The other three SBRs were operated according to the BPR principle and an anaerobic/aerobic sequence was used for each reactor. Effluent phosphorus concentrations of the BPR reactors were significantly lower than that of the fully aerobic reactor and effluent soluble phosphorus concentrations below 0.3 mg/L were achieved. Other major effluent parameters such as BOD5, COD, AOX, pH of the BPR reactors were similar to those of the AS reactor. Moreover, values of these parameters were very close to those of the actual pulp mill AS wastewater treatment plant from which untreated wastewater was collected. This lab-scale study clearly demonstrated the potential for the application of the BPR process for the treatment of kraft pulp mill wastewater.

Keywords: Activated sludge process, biological phosphorus removal, pulp mill wastewater, sequencing batch reactor

Sulfidogenic anaerobic treatment to remove organic matter, sulfate and AOX from Kraft pulp mill wastewater

M. Rissato[1], D. Botelho[1], A. Rodrigues[1], E. Foresti[1] and E. Cleto[1]

[1]*Laboratório de Processos Biológicos, Escola de Engenharia de São Carlos, Universidade de São Paulo. Av. Trabalhador São-carlense, 400, 13566-590, São Carlos, SP, Brazil. *e-mail: ecpires@sc.usp.br*

One obstacle to increase water recycle is the contaminant buildup that occurs as the recycle ratio increases, and sulfur compounds are especially detrimental. Sulfate removal can be achieved sulfidogenic anaerobic microorganisms to reduce the sulfate using organic matter as electron donors. This work was designed to assess the behavior of a sulfidogenic reactor treating wastewater from a kraft pulp mill. A fixed bed horizontal anaerobic reactor (100 cm of length and 5 cm in diameter), operating at 30 + 1°C and hydraulic detention time of 12 + 1h was used to treat the effluent with the following characteristics: COD: 2270 + 75 mg/L; sulfate: 500 + 55 mg/L and, AOX: 24.2 + 1.5 mg/L. After sulfidogenesis was established the reactor showed steady performance removing 60.1 + 2.6% of the COD; 98.1 + 3.9% of the sulfate and 73.1 + 3.6% of the AOX. Concentration profiles along the reactor length showed that volatile fatty acids, AOX and sulfate were removed in the first one third of the reactor. However, the removal of the remaining biodegradable COD occurred up to the exit section. The AOX and COD removal efficiencies observed in the current work are higher than the ones found by other researchers. An explanation for this performance may rest on the sulfidogenic path that efficiently used the available organic matter for reduction of the sulfate to sulfite. To finish the treatment it is necessary to remove the sulfite, which is easier to be removed than sulfate.

Keywords: anaerobic digestion; pulp mill; wastewater recycling; fixed film reactor; sulfate

Electrocoagulation of a paperboard mill wastewater using varying levels of tin oxide coating on aluminum electrodes

M. Mokua[1], O. Orori[1], M. Mwamburi[2], K. Balozi[1], K. Senelwa[1] and L. Etiégni[1]*

[1]*Department of Forestry and Wood Science, Chepkoilel University College, Box 1125 Eldoret, Kenya.*
[2]*Department of Physics, Chepkoilel University College, Box 1125 Eldoret, Kenya*
**Corresponding and presenting author: lazetiegni@amatala.org*

Electrocoagulation is a viable option for industrial wastewater treatment, but depends heavily on the type of electrodes used. Graphite, aluminum and iron have been some of the electrode types used to treat wastewater. Unfortunately these electrodes wear out too often due to corrosion at the anode and/or passivation at the cathode. A layer of tin oxide on aluminum electrodes could provide a good protective layer thereby extending their life. An experiment was carried out to determine the effect of increasing the level of tin oxide coating on aluminum electrodes during electrocoagulation of paperboard factory wastewater. Ten levels of tin oxide coating were considered. Physicochemical properties of the wastewater were determined before and after treatment. Power intensity to a predetermined endpoint was also calculated for every level of coating. Data were collected through a systematic sampling design and analyzed using the Statistical Package for Social Scientist (SPSS) Version 16.0. Post hoc tests were run to separate various means. The results showed that an increase in the level of electrode coating led to a decrease in power intensity from 1.8% to 18.2%. Biochemical Oxygen Demand decreased from 37.5% to 93.7% depending on level of tin oxide coating. Similarly total solids and total suspended solids decreased from 23.9% to54.0%, and from 43% to 62% respectively with the level of electrode coating. It was therefore recommend that an electrolyte be used during electrocoagulation method for further reduction of power intensity and economic considerations determine the optimum coating for the paperboard wastewater treatment.

Keywords: aluminum; electrocoagulation; electrodes; paperboard wastewater; tin oxide

The study was undertaken to determine the effective level of tin oxide coating on aluminum electrodes.

RESOURCE AND ENERGY RECOVERY OR CONSERVATION

Polyhydroxyalkanoates production from wood mill effluents

M. Vila, T. Mato, C. Kennes and M.C. Veiga*

Chemical Engineering Laboratory, Faculty of Sciences, University of A Coruña, Rúa da Fraga, nº 10, 15008 A Coruña, Spain
**Corresponding author: veiga@udc.es*

Fibreboard production is one of the most important industrial activities in Galicia (Spain). The objective of this work was to study the viability of producing biodegradable plastics (PHA) using the wastewater generated in this type of industry. The PHA production process was developed in three different steps.
In the first step, the wastewater was fermented in a sequential batch reactor in order to convert the complex organic matter into low chain fatty acids, which would be used in a second step as substrate for PHA production. The effect of the pH on the acidification process was studied, obtaining the highest acidification at pH 5.5, remaining 27% of the COD not degraded.
In the second step, a mixed culture was enriched in bacteria with high storage ability in a sequencing batch reactor operated under aerobic dynamic feeding (ADF) conditions and fed with the fermented wood mill effluent. The effect of the suspended solids present in the wastewater was studied, applying the following operational conditions: TRH 1d, TRS 10d, 30Cmmol/l of carbon source and 1Nmmol/l of nitrogen. A higher PHA accumulation rate was obtained in the effluent without suspended solids, 40% of weight of dry cells. However, the suspended solids concentration did not affect the storage yield, 0.30-0.40 Cmmol HA/ Cmmol AGV in both cases.
Once the reactor was in a stationary phase, several fedbatchs were developed to determine the storage capacity and the storages yields of the sludges, being slighted higher the storage yield when the suspended solids were present.

Keywords: Acidogenesis, Volatile fatty acids (VFA), polyhydroxyalkanoates (PHA), Sequencing batch reactor (SBR), Aerobic dynamic feeding (ADF) conditions, wood mill effluents

Effect of hydrolysing paper and pulp fibre sludge with enzymes on biogas production in anaerobic batch reactors

M. Badshah, W. Parawira and B. Mattiasson

Department of Biotechnology Lund University. P. O. Box 124, SE-221 00 Lund, Sweden.
E-mail: Malik.Badshah@biotek.lu.se

Addition of enzymes was done for enhancement of biogas production from paper and pulp mill sludges. Enzymes used were AccelleraseTM 1000, GC 140 xylanase and proteases, alone and in different combinations. The fiber sludge was evaluated for simultaneous enzymatic treatment and anaerobic digestion using combinations of Accelerase and xylanases. Also biogas production from the fiber sludge was evaluated by using different sources of anaerobic inoculum. During enzymatic pretreatment and anaerobic digestion the fibre sludge and sludge containing a thickener from the paper and pulp mill were kept at 37 °C and 70 rpm shaking.

The methane yield of fibre sludge of untreated and treated sludge with enzymes using an inoculum from a digester fed with pig manure, slaughterhouse waste and waste from a food processing industry, in ratio of 1:2 in terms of VS, 95-98% of theoretical methane yield was achieved (assuming the fibre sludge as cellulose fibres). However, sludge from a municipal wastewater treatment plant gave 80-90% of theoretical yield. Also the VS reduction ranged from 94-100% in different scenarios.

The methane production from sludge with thickener was during18 days increased by 48±26 % after the addition of AccelleraseTM 1000. Similarly, methane production from same sludge during 16 days was increased by 17 ⊥ 11% with the pretreatment with xylanase. Proteases were added when methane production from the sludges treated by xylanase and Accelerase ceased. This¬ resulted in increase of 10 ±4 and 2.2 ± 0.6 ml/ g VS added for the sludge pretreated with Accellerase and xylanase respectively.

Keywords: Anaerobic digestion, Enzymatic pretreatment, pulp mill sludges, Biogas

Chemical flocculation as pretreatment for energy-efficient biological treatment

M. Sandberg and A. Rixen

Karlstad University, Faculty of Science and Technology, Department of Energy, Environmental and Building Technology, Karlstad, Sweden, email: maria.sandberg@kau.se

Biological treatment of forest industry wastewaters uses large amounts of electrical energy. For many mills aeration is the single most energy intensive process in the wastewater treatment plant. The oxygen transfer from bubbles into the water volume is quite slow with low aeration efficiency as result. If the oxygen transfer can be improved, the efficiency can be improved and less compressed air will be needed for aeration with decreased energy costs as the result. It has been shown that surface-active extractives, such as fatty acids and resin acids, have a large impact on oxygen transfer in diffuser aeration systems. With chemical flocculation, extractives can be removed from the process streams. Trials have been made where pulp-mill process streams were pretreated with chemical flocculation before the oxygen transfer tests. When the chemical flocculation conditions were optimized to increase surface tension in the process stream, extractives were removed and the oxygen transfer rate increased in the following oxygen transfer tests. With the pretreatment also COD concentration decreased. The effect will be less need for oxygen for degradation in the following biological process. With decreased COD concentration and increased oxygen-transfer rate, energy can be saved in the biological treatment. Energy savings resulted in greater economic return than the cost for flocculating chemicals. If waste sludge can be used as energy source, one can even achieve a better result.

Acknowledgements. This work was supported by ÅF-forskningsstiftelse and Stora Enso Skoghall mill.

Keywords: aerobic treatment, pulp-mill effluent, extractives, oxygen transfer, chemical flocculation

Process integration of effluent treatment techniques – a way of decreasing the energy and resource use in pulp and paper industries

T. Ericsson, Å. Sivard and H. Norrström

ÅF, Frösundaleden 2, S-169 99 Stockholm, Sweden Ph +46 10 505 1307, fax +46 10 505 0010,
E-mail: tomas.j.ericsson@afconsult.com, asa.sivard@afconsult.com, hans.norrstrom@afconsult.com

The operating costs for waste water treatment in pulp and paper mills are high and the treatment stages have so far only to a minor extent been integrated in the mill's total process systems. In a desk study it has been shown that process integrated methods could be cost and energy efficient ways to reduce the environmental impact.

Anaerobic treatment and membrane methods combined in different treatment sequences were used for treatment of effluents from Kraft pulp, CTMP, TMP and RP mills with and without integrated paper production. The production capacities, effluent volumes and discharge levels were chosen to be representative for Scandinavian mills. Membrane filtered water (mainly of high temperature) was reused in the process and accordingly the fresh water consumption was reduced.

The process integrated cases were compared with conventional processes including waste water treatment in an activated sludge plant. Part of the existing effluent treatment plant was used as polishing stage.

The process integrated cases proved to be superior in almost all aspects. The Electric power use was reduced by 15-30 MWh/day, the cooling demand by up to 8 500 MWh/day and 35-90 MWh/day biogas could be generated. It must be stressed that the treatment and the mill concepts are simplified examples and cost calculations are based on generalised assumptions on capacity, treatment efficiency etc. Considering the encouraging results it is suggested that process integration approaches could lead to more resource and cost efficient alternatives compared to present effluent treatment plants.

Keywords: anaerobic treatment, membrane methods, pulp &paper mill effluent, reduced power use, biogas production, water reuse

FATE AND EFFECT OF SPECIFIC COMPOUNDS DURING WASTEWATER TREATMENT

Chlorate behaviour in Kraft pulp wastewater treatment systems

M. Cabrera[1], S. Toivakainen[2], M. Martikka[2] and O. Dahl[2]

[1]*Forest Process Engineering Group- Engineering School- Universidad de la República- Uruguay. Email: ncabrera@fing.edu.uy.* [2]*Clean Technologies Group- Forest Products Technology Department- Aalto University-Finland*

Nowadays, the only major source of chlorate input to the environment is effluent from pulp mills using chlorine dioxide bleaching. Chlorate discharges to the environment are important to control due to they showed some toxicity to certain algae species and could change the aquatic ecosystem. In this work, chlorate behaviour in anaerobic conditions was studied in laboratory reactors. The results show that chlorate could be completely reduced in anaerobic condition in 4 hours, however neither COD reduction nor methane formation could achieve in this reactors. Simultaneously chlorate, ORP, dissolved oxygen and other parameters are measured in several parts of the wastewater treatment system of three Kraft pulp mills with different aerobic treatment configurations. In these mills chlorate is completely removed in the wastewater treatment system. In parts of the system where low oxygen content and reductive media conditions are reached (i.e. equalization basins, sedimentation basins) chlorate is eliminated, probably because it could be used by some microorganisms as an electron acceptor. Consequently no particular treatments are needed to remove chlorate; the conditions achieved in the stabilization ponds in the modern mills are enough to complete chlorate reduction.

Keywords: Chlorate removal, ORP, Kraft pulp mills

Organic load removal mechanisms in a eucalyptus bleached kraft pulp mill activated sludge plant

C. Lucio da Silva, A. Mounteer, T. Duarte, S. Pereira and G. Mozer.

Department of Civil Engineering, Federal University of Viçosa, Brazil

The relative importance of adsorption and degradation for removal of organic load during biological treatment of bleached kraft pulp mill effluent was investigated in a mill activated sludge plant and in laboratory assays. Samples of primary and secondary effluents and return activated sludge were collected monthly over five months at a Brazilian eucalyptus bleached kraft pulp mill activated sludge plant. Effluents were characterized by quantifying dissolved organic carbon (DOC), chemical (COD) and biological (BOD) oxygen demands, adsorbable organic halogen (AOX), residual lignin, lipophilic (dichloromethane) extractives and specific ultraviolet absorbance (SUVA254). Extracellular polymeric substances (EPS) extracted from the sludge samples were characterized in terms of volatile suspended solids (VSS), protein, carbohydrates and DOC. Effluent samples and EPS extracts were analyzed by GC-MS to identify the main compounds present in each sample type. In addition, a laboratory study was conducted to assess the relative importance of degradation and adsorption by biomass in aerobic biological treatment, using subsamples of the primary effluent and activated sludge samples collected monthly. The effluent was as inoculated with activated sludge or chemically treated activated sludge (1.25 g sodium azide/g SSV, 30 min, 100 rpm) and incubated for 24h. The chemical treatment was performed to reduce microbiological activity, but avoid bacterial cell lysis and EPS degradation. After the incubation period, effluent COD, DOC, residual lignin and extractives were quantified. Uninoculated effluent was used as control to verify the role of volatilization in removal of organic matter. Mill effluent characterization indicated removal efficiencies of 85% of BOD5, 78% of DOC, 40% of COD, 38% of AOX and 17% of residual lignin. Lipophilic extractives increased by 23% on average, paralleled by a 26% increase in the SUVA254. EPS from the mill treatment plant sludge had an organic carbon content of 6.0 mg/g VSS with 3.5 times more proteins (19.9 mg/g VSS) than carbohydrates (6.1 mg/g VSS). The six main classes of compounds detected in primary effluent samples, in decreasing order, were fatty acids, sterols, carbohydrates, hydrocarbons, and alcohols. Only fatty acids, sterols and hydrocarbons were found in the secondary effluent samples, in much smaller quantity than in the primary effluent (total relative chromatogram peak area of 5.2 versus 16.2 in primary effluent). EPS extracts contained, in decreasing order, fatty acids, sterols, hydrocarbons, alcohols and amino acids. Octadeca-9,12-dienoic and hexadecanoic acids, recognized as Eucalyptus spp. extractives and two unidentified sterols were found in greater relative amounts in EPS extracts than in secondary effluent while octadecan-1-ol was only found in primary effluent and EPS extracts. These results indicate the importance EPS adsorption for removal of these lipophilic wood extractives from pulp mill effluent. In the laboratory study, COD, DOC, residual lignin and extractives values were 10 to 50% lower in effluents treated with active biomass than with chemically inactivated biomass. Volatilization appeared to play an important role in removal of DOC and COD. EPS. EPS extracts of the activated sludge after laboratory incubation had higher organic carbon contents per gram of suspended solids than the mill activated sludge samples with protein/carbohydrate ratios of 5.65 for active biomass and 6.95 for inactive biomass, suggesting that laboratory induced stress led to greater EPS production. The difference in effluent organic content after treatment with active and chemically treated biomass serves as an estimate of the amount of organic matter that would adsorbed and thus removed with the excess sludge wasted in the mill treatment plant. Knowledge of the type and quantity of recalcitrant organic matter adsorbed on the excess sludge, such as the lipophilic Eucalyptus extractives found in the EPS extracts in this study, is important to determine the risks associated with alternative sludge management practices

Towards understanding colour behaviour in pulp and paper mill wastewater treatment systems

G. Gielen, K. McGrouther, D. Gapes and T. Stuthridge

Scion, Private Bag 3020, Rotorua 3074, New Zealand.

Increased awareness regarding aesthetic impacts of pulp and paper mill wastewaters have resulted in increased regulatory restrictions around emission of coloured material via these wastewaters. Meeting such discharge restrictions now and into the future will require cost-effective approaches to colour reduction prior to final discharge. To achieve this, it is vital that we increase our understanding of colour behaviour in pulp mill wastewater treatment systems.

The focus of this study was to observe the impact of a major disruption to influent characteristics at a full-scale pulp and paper mill treatment system. The lagoon system under investigation received effluent from a CTMP mechanical pulp mill, Kraft pulp mill, tissue mill, and paper mill. A planned two-month Kraft mill shut, provided a unique opportunity to observe the effect of factors such as organic load, pulping source, hydraulic retention time, and sulphur levels on colour behaviour in the treatment system. To investigate correlations between system conditions and the observed colour behaviour, samples were taken throughout the wastewater treatment system during and after the Kraft mill shut.

This paper will discuss the relevant correlations which were found, and relate them to findings from the wider research literature and previous studies from our laboratory and fieldwork. Conditions which have been associated with colour and colour formation will be discussed in relation to the behaviour encountered within the investigated treatment system.

Concluding remarks on the opportunities and challenges in colour minimisation for pulp and paper mills will be presented within the context of this study.

Keywords: colour, Kraft, CTMP, colour formation, pulp and paper wastewater, influent variability

Influence of nonionic surfactants and biocides on the anaerobic stages of paper mill waste water treatment plants

A. Klose[1], T. Hoffmann[2] and E. Tiedtke[3]

[1]*Chair of paper technology and mechanical process engineering – Germany, Technische Universität Darmstadt, E-mail: klose@papier.tu-darmstadt.de.* [2]*Department of chemistry and biotechnology – Germany, Hochschule Darmstadt E-mail: tobias.hoffmann@stud.h-da.de..* [3]*Kolb distribution Ltd. – Switzerland, Hedingen, E-mail: Elke.Tiedtke@kolb.ch*

The majority of German paper mills utilize recovered paper as raw material. Recovered paper is known as perfect inoculum to process water cycles of paper production. To prevent microbial induced impacts on runability and paper quality a wide variety of nonionic surfactants and biocides is commonly used. Most additives have been specially developed and customized for application in these process water systems. However, only little information concerning the influences of those additives on anaerobic waste water treatment stages is known. Furthermore some observation in industrial use seemed to support the idea that nonionic surfactants affect the performance of anaerobic waste water plants.

Aim of this work was to determine the impact of different additives (volume enhancers, biocides, deposit dispersants) on anaerobic digestion. Both batch and continuous trials were carried out. Batch trials took place in glass bottles (1 L) with a corresponding biogas measuring device on top. Continuous trials were carried out in UASB reactors (Plexiglas®, 30 L), that were fed in bypass to an industrial plant. Biogas production, COD degradation, acid concentration, pH-values and temperature were measured regularly.

The batch tests showed that all studied additives only had very little influence on system stability and biogas production in intended concentrations. For very high concentrations some impacts on COD degradation and biogas production were detected. To cover long term effects of additive dosages into an anaerobic treatment stage the continuous trials were carried out. Comparison with a reference system did not show any negative influences of dosage in intended concentrations.

Keywords: anaerobic treatment, biocides, non-ionic surfactants, batch test, biodegradability, UASB-pilot plant

Biological activity of bleached kraft pulp mill effluents before and after activated sludge and ozone treatments

A. Cunha, A. Mounteer and T. Valverde

Civil Engineering Department, Federal University of Viçosa, Viçosa, MG, Brazil

The pulp industry, one of the most important sectors of the Brazilian economy, is responsible for generating large effluent volumes containing high organic loads, part of which is recalcitrant to conventional biological treatment. Studies with model ecosystems have indicated a direct correlation between organic matter remaining after biological treatment and biological activity of effluents in the aquatic environment, with toxicity and endocrine disruption important forms of expression of this biological activity in aquatic organisms present in receiving waters. The objective of the present study was to quantify the biological activity in beached kraft pulp mill effluent using acute and chronic toxicity and endocrine disruption assays and to evaluate the effect of ozonation as tertiary treatment for removal of the biological activity. To this end, primary and secondary effluents were collected bimonthly over the course of one year at a Brazilian bleached eucalypt kraft pulp mill and characterized by quantifying BOD, COD, TOC, AOX and residual lignin. The nature of the effluent components responsible for secondary effluent chronic toxicity and estrogenic activity were investigated using the USEPA's toxicity identification evaluation (TIE) phase I protocols. Secondary effluent was also treated with ozone (50 mg/L) in a bench-scale reactor. The mill effluent treatment plant presented average removal efficiencies of 91% for BOD (from 477 to 45 mg/L), 72% for COD (from 956 to 263 mg/L), 76% for COT (from 231 to 56 mg/L), 50% for AOX (from 3 to 1.5 mg/L) and 65% for residual lignin (from 57 to 20 mg/L) over the sampling period. The primary effluent samples presented no acute toxicity to the microcrustacean Daphnia similis, but presented chronic toxicity to the green alga Pseudokirchneriella subcapitata (lowest observable effect concentration, LOEC = 12.5%) and the microcrustacean Ceriodaphnia dubia (LOEC = 9 ±4%). In general, chronic toxicity was reduced during biological treatment in the mill´s activated sludge wastewater treatment plant, although the secondary effluent still presented chronic toxicity (LOEC = 20 ±7% for P. subcapitata and 12.5% for C. dubia). Toxicity was higher in the high molar mass secondary effluent fractions (> 500 g/mol) that contained, on average 87% of the COD, 70% of the TOC and 75% of the residual lignin. On some collection dates, the effluent dilution necessary to eliminate the chronic toxicity effect was higher than the dilution occurring through effluent discharge to the receiving waters. The primary effluents also presented estrogenic activity, as quantified in the yeast estrogen screen (YES) assay, with some removal achieved during biological treatment (10.3 ± 3 μg/L EQ-E2 in primary effluent and 7.6 ± 1.5 μg/L EQ-E2 in secondary effluent). The TIE effluent manipulations that reduced effluent chronic toxicity to both C. dubia and P. subcapitata were solid phase extraction, aeration and pH adjustment to pH 3, whereas only solid phase extraction reduced (but did not eliminate) estrogenic activity. Ozone treatment of the secondary effluent resulted in additional average removals of 32% COD, 20% TOC and 75% lignin, but had no effect on BOD. Ozone treatment eliminated chronic toxicity to C. dubia and P. subcapitata and reduced estrogenic activity below the assay detection level. The results suggest that Brazilian pulp mills may be releasing biologically active compounds at ecologically relevant concentrations to receiving waters and that an alternative to minimize this risk is tertiary ozone treatment.

Abatement of 4-chloropenol under visible light with peroxymonosulfate (PMS) in presence of Co^{2+}

J. Fernandez[1] and N. Diaz

Catholic University of the Santísima Concepcion, Faculty of Sciences, Alonso de Ribera 2850, casilla 297 Concepción, Chile. e-mail: jfernandez@ucsc.cl

The 4-chlorophenol is a highly toxic organic compound present in wastewater, mainly from brokers or chemical byproducts of the petrochemical, paper, plastics and conventional disinfection of drinking water. In the present study evaluated a similar treatment of advanced oxidation technologies involving the degradation of 4-chlorophenol (4-Cl) by peroxymonosulfate (PMS) catalyzed by cobalt (Co+2) and assisted by visible light.

To achieve the objective previous degradations were conducted to in order to set the intervals to conduct a multivariate analysis. According to these results experimental design was carried out between the ranges of 50 - 100 mg L-1 for 4-Cl, 3,88 - 7,76 mM for PMS and 0,5 - 1 mg L-1 for Co+2, the response factor was the residual 4-Cl concentration after 60 min of irradiation. The design was performed and analyzed using the statistical program MODDE 7.0, resulting in a star-like array of 17 experiments.

The polynomial expression and the response surface obtained gave an optimal response study ([4-Cl] = 50 mg L-1 , [PMS] = 7,76 mM and [Co+2] = 1 mg L-1). Under that conditions a total TOC reduction (22 to 0 mg L-1) after 300 minutes, also total COD removal of 124 (0 min) to 0 mg L-1 (300min) and chloride ions 1,7 (60 min) to 0 mg L-1 (120 min). HPLC-MS experiments were conducted to identify degradation intermediates.

Keywords: 4-chlorophenols, oxone, AOPs, PMS, HPLC-MS

MANAGEMENT OF WASTE SOLIDS AND THEIR EVALUATION AS BY-PRODUCTS

Waste management: a review of the state of art of pulp and paper mills in South America

A. Rezende[1], M. Diocaretz[2], J. Castañeda[1,3], M. Grandon[2] and G Vidal[2]

[1]Department of Civil Engineering - Federal University of Viçosa – Viçosa-MG/Brazil ana.augusta@ufv.br. [2]Environmental Science Center EULA – Chile, University of Concepción. P.O. Box 160- C. Concepción-Chile. E-mail: glvidal@udec.cl. [3]Faculty de Engineering – University San Carlos of Guatemala – Guatemala

Waste management has become a key issue in the pulp and paper mill. This industry is disbursing high cost in waste landfill disposal, because of the increasing generation of wastes, such as primary and secondary sludge from wastewater treatment, dregs and grits from caustification process and ashes from power boilers. Moreover, the increasingly stringent environmental regulations and marketing demands have led the industry to find more sustainable solutions than landfilling. The pulp and paper production is increasing in South America with the construction of new several mills in Brazil, Chile and Uruguay which are incorporating alternatives for improving the waste management. Options such as combustion for steam and energy production, pyrolysis, gasification, composting, land spreading, and reuse as building material are being studied and applied. This review aims to present the state of art of pulp and paper mills waste management in South America. The main characteristics and management aspects of this industry wastes are described, including minimization, treatment and reuse processes. Also, a summary of the regulations and the main research results reported in the literature are provided.

Acknowledgements. Dr. Rezende thanks to Federal University of Viçosa for supporting his stay at the Environmental Science Center EULA-Chile, University of Concepción.

Keywords: pulp & paper mill, waste management, solid waste treatment, pulp residues

Co-firing of biological sludge and wooden residues from pulp mills for energy production

V. Rodrigues[1], C. Mudado[1], A. Carneiro[1], A. Rezende[2], W. Quirino[3], L. Santos[1] and G. Ikawa[1]

[1] *Departamento de Engenharia Florestal - Universidade Federal de Viçosa Viçosa-MG/Brasil mudado@ufv.br.* [2]*Departamento de Engenharia Civil - Universidade Federal de Viçosa Viçosa-MG/Brasil.* [3]*Laboratório de Resíduos Florestais – IBAMA, Brasília-DF/Brasil*

Pulp mill generate large amounts of solid waste, particularly wood residues and sludge from effluent treatment plants. Aerobic activated sludge is the most common effluent treatment process used worldwide. In this process, bio-sludge generation can be as high as 20 kg per ton of pulp produced (dry basis) and normally it has been disposed in landfills. Incineration and energy production are becoming the main waste recovery method for pulp mills. Currently, in the kraft pulping process, the energy is generated in the chemical recovery boiler and in the biomass boiler, with the combustion of black liquor and wood residues, respectively.

Bio-sludge contains a high amount of organic matter which provides a potential for energy recovery from this material. Recently, co-firing of wood residuals and other solid wastes such as sludge, municipal solid waste and biomass has been considered an environmentally sound, and an interesting economic approach to both waste remediation and energy production. However, co-firing process of biomass and pulp mill bio-sludge has not been widely studied and its environmental impact is still uncertain.

In this paper, it was studied the technical feasibility of co-processing the biological sludge and wood residues through combustion process of a Brazilian kraft pulp mill and it was investigated the major environmental aspects related to air emissions and ashes quality. The biological sludge, wood residues and ash were classified as non hazardous. The combustion of the mixture of 20% bio-sludge and 80% wood residues did not present emissions above the legislation limits for boilers with nominal power over 70 MW.

Keywords: pulp mill, biological sludge, energy recovery, wood residues, waste co-processing

Minimisation of hazardous biosludge from pulp and paper industry by ozone oxidation

S. Gupta[1], S. K. Chakrabarti[2], Satnam Singh[1]

[1]*School of Chemistry and Biochemistry, Thapar University, Patiala – 147004, Punjab, India. sanju_chem@yahoo.com (S. Gupta), ssingh@thapar.edu (Satnam Singh).* [2]*Thapar Centre for Industrial Research & Development, Paper Mill Campus, Yamuna Nagar- 135001, Haryana, India. swapan@thaparpaperresearch.org, Phone: +911732292703, FAX: +911732292748*

The wastewater from integrated pulp and paper mill contains a high concentration of organochlorine compounds and a part of these compounds is adsorbed on biomass during biological treatment by activated sludge process (ASP). Due to presence of organochlorine compounds, the biosludge from pulp and paper industry has been classified as hazardous waste in India, and handling and disposal of biosludge therefore is a cause of concern. The study is aimed to establish the feasibility of disintegration of biosludge with ozone and subsequent biodegradation of disintegrated biomass in the ASP. 16.0±0.5% solubilisation of MLVSS content and 22.6±2.3% mineralization of AOX compounds were observed at ozone dosage of 43-47 mg O3/g dry solid (DS). Solubilisation of biomass resulted in increase in more biodegradable COD. Three times of excess sludge from continuous reactor was ozonated daily at average ozone dosage of 45.1 mg O3/g of DS and recycled back to reactor. The performance of ASP was better than that in control reactor in term of COD, TOC, colour and AOX reduction. Ozone treatment controlled the proliferation of filamentous organisms and there was drastic change in sludge volume index (SVI) due to decrease in surface charge. The index dropped to 67±34 through ozone treatment whereas it was 340±33 ml/g in control reactor. The disposable biomass through ozone treatment process was 1/5 of that in the control reactor. The recycling of ozone treated sludge did not affect the AOX concentration in the biosludge.

Keywords: Activated sludge process, biosludge generation, AOX in biosludge, ozone treatment, sludge volume index

Investigation of thermal, physical and chemical characteristics of solid waste from cellulose pulp industry

L. de Caux, L. Dalvi and M. Fonseca

Celulose Nipo-Brasileira-CENIBRA S/A, BR 381 Km 172, Belo Oriente MG, Brasil,
Email: leonardo.caux@cenibra.com.br

Sustainable natural resources use and better waste management have acquired a strategic connotation to industries due to restrictive legislation evolution, large waste generation and rapid natural sources depletion besides the processes energy optimization necessity. Marketable products creation from solid wastes has been highlighted as a great management alternative. However, deep knowledge of its characteristics is needed to appropriate destination. Water treatment station sludge (SWTS) is included in this context and traditional techniques used are: composting, soil conditioning, power generation, road paving and construction use. Aiming to verify the potential for waste reuse from the water treatment station of CENIBRA S/A were evaluated its characteristics by X-ray diffraction, X-ray fluorescence, atomic absorption spectrometry, infrared spectroscopy with Fourier transform, thermogravimetric analysis and differential thermal analysis. Also are presented two cases of waste management. First one is focused on sintered red ceramics production and second one focused on raw materials of water treatment recovery. Characterization results indicated clay minerals presence associated with water treatment chemicals. Results also showed that SWTS has some ceramic formulations problems among them the volumetric shrinkage due to large amount of ferric oxide and porosity causing resistance reduction. However, was observed that the intrinsic resistance of waste ceramic bodies is similar of those made by clay. Flocculants recovery showed great potential, not only due to chemicals but also the sludge leaching producing a new waste with a lower aluminum and iron content, contributing to an improvement on ceramic formulations and raising the usage possibilities, besides the significant water recovery.

Keywords: Solid waste, thermal characteristics, chemical characteristics, red ceramics, chemical recovery

Microwave-assisted chemical activation of pulp mill sludge with potassium hydroxide for endocrine disrupting compounds removal

A. Namazi, D. Allen and C. Jia

Department of Chemical Engineering and Applied Chemistry, 200 College St, Toronto, ON, Canada, M5S 3E5
A.bagherzadehnamaazi@utoronto.ca

Using a microwave oven, activated carbons were produced from various pulp mill sludges via chemical activation with potassium hydroxide (KOH) as an activating agent. Microwave-assisted KOH-activation was carried out under various activation times (3-10 min) and KOH/carbon mass ratios (1/1 - 2/1). The activated carbons produced were characterized using an elemental analyzer, a surface area analyzer, and a scanning electron microscope. The results were compared with those of activation with the electric furnace at 400 - 600 ºC for 1- 2 hours. The effectiveness of the activated carbon in removing endocrine disrupting compounds (EDCs) was tested. The results show that pulp mill sludge is a suitable precursor of activated carbon. The use of microwave substantially reduces the activation time. The performance of the pulp-mill sludge-derived activated carbon is similar to that of a coconut shell-derived commercial activated carbon.

Effects of the successive application of stabilized secondary sludge from the cellulose wastewater treatment plant to degraded chilean volcanic soils

F. Gallardo, M. Flores and M. Diez

Scientific and Technological Bioresource Nucleus-BIOREN, Universidad de La Frontera, Temuco- Chile.
E-mail: fgallar@ufro.cl

The aim of this work was to evaluate the effects of the successive application of stabilized secondary sludge from cellulose on the physico-chemical properties of soils of southern Chile. The soils used in this study were two Andisol belonging to Gorbea and Freire series and an Ultisol (Collipulli series) characterized by different contents of organic matter and pH. The sludge used as amendment was stabilized pulp mill sludge (pH 7.0; OM 75%; N 586 mg/kg; P 313 mg/kg) obtained from an aerated pond that treated bleached kraft mill wastewater collected from a landfill after one year disposal. The sludge application in the soils was realized at rates of 0, 10, 20 and 30 t/ha divided in four applications during one year. The effect of the addition of sludge on the soil was evaluated through macro and micro nutrients soil levels. The application of different doses of pulp mill sludge in the soil improved the physico-chemical properties of soils, especially after the third application. In general, pulp mill sludge application increased the contents of O.M, N, P, macro and microelement content. Specifically, the Ultisol soil belong Collipulli series showed the largest increase in O.M content (27%). This would indicate that soils with lower organic matter content are more likely to be rewarded by the application of a highly organic amendment such as pulp mill sludge.

Acknowledgements: This work was supported by Fondecyt 1080427

Keywords: Degraded chilean volcanic soils, Pulp mill sludge, amended soil

International Water Association | 10th IWA Symposium on Forest Industry Wastewaters

POSTER PRESENTATIONS

LIFE CYCLE ASSESSMENT AND ECOLOGICAL FOOTPRINTS (WATER AND CARBON FOOTPRINT)

NORSKE SKOG Biobío water management proposal based on a Water Footprint approach

S. Sanzana, C. Ulloa and D. Caamaño

Environmental Science Center EULA – Chile, University of Concepción. P.O. Box 160- C. Concepción-Chile. E-mail: claudiaulloa@udec.cl.

The present work utilized the water footprint concept, applied to the product, to quantify the volumes of used freshwater during the paper production of the Norske Skog Biobío pulp mill located near the mouth of the Biobío river, VIII Region – Chile.

The water footprint was calculated to one line of production, starting from the wood preparation in the pulp mill facilities up to the final product. Both, the gray and the blue components were considered, and critical points of water consumption were identified.

The highest value of the water footprint was determined and equivalent to 5600 m^3/s, and it takes place in the paper machine. This hot spot corresponds to 50% of the total water footprint value. Management actions were proposed to decrease the usage of water in this critical point. Results indicated that a 58% increment in the volume of recirculated water decreased the water the footprint in a 48%.

The variation of the water footprint under two new scenarios was also studied. The first scenario supposed an emissary placing the effluent into the Pacific Ocean directly, and the second situation considered the natural pollutant concentrations in the river changed with time and space. The first condition reduced 50% the blue water footprint, and the second scenario caused a diminution of the 60% in the gray water footprint component. Concluding that the analysis is site specific, and strongly depended on the defined river natural conditions - suggesting a high component of seasonal fluctuation for the water footprint output.
Acknowledgements: The authors wish to acknowledge the Biobío Region Cleaner Production Council for funding this research, and also, Norske Skog Biobío for granting access to the facilities and for the valuable information provided.

Keywords: Water Footprint, paper mill effluent, water use optimization

Effect of process modifications on the chilean cellulose Water Footprint: a life cycle perspective

P. González[1], M. Vega[2]

[1]*Department of Public Health, Faculty of Medicine, University of Concepcion. Email: pagonza@udec.cl.*
[2]*Chemical Engineering Department, Engineering Faulty, University of Concepción. Email: mabvega@udec.cl*

Environmental loads associated to cellulose production spread well beyond the mill gate. In particular, the potential impacts of such industrial activity on water resources are found all through the cellulose product life cycle, including plantations, transport, chemicals manufacturing and cellulose making. In Chile, around 40% plantations are located in the Biobio river basin, accounting for more than 40% total national cellulose production. Increasing environmental pressures have led to the implementation of substantial process improvements, reducing specific water usage in most mills. Indeed, average usage dropped from 45 m3/ton to 70 m3/ton cellulose from 1998 to 2008. Also, there have been significant changes in bleaching operations, and effluent treatment systems. However, despite these improvements there is still concern about the implication of these changes on water resources. Climate change trends indicate that water availability in the Biobio river basin is steadily decreasing posing a potential threat to forestry and agricultural activity in this Region. In this context, this paper presents results on the effect of technological changes on the water footprint associated to cellulose production, using a life cycle assessment (LCA) approach.

LCA is a powerful tool for environmental management and process design, since it allows a systemic assessment of environmental loads along all processes comprising the product life cycle. This study focuses on the Biobio river basin production system, and primary production data was obtained from 1994-1996 and 2007-2009 as reference periods. A cradle to gate approach, considering nurseries and plantations as cradle, and port facilities as the gate, was used here. Biobio river secondary quality standards were used as reference standards to calculate grey water footprint.

Results show that plantations are responsible for most green water, due to the high rates of evapo-transpiration, whereas industrial operations contribute with most blue and green water. Moreover, there are significant differences in water footprint associated to eucalyptus and pine pulp, particularly due to differences in wood density, growth rate and process yields.

Although specific water usage has been reduced due to process improvements, green and blue water consumptions was little affected by such technological changes. On the other hand, grey water showed over 80% reductions as a result of significant decreases in COD, and AOX loads, due to new effluent treatment systems. The implications of these findings are discussed in the context of climate change trends in the Biobio river basin.

Keywords: cellulose, environmental impact assessment, life cycle, process modifications

INTERNAL CONTROL AND SYSTEM CLOSURE

Qualitative and quantitative aspects consumption of water and wastewater generation in pulp and paper mill

C. Xavier[1], S. Vanzetto[1], M. Klenk[1], T. Oliveira de Melo[1] and G. Joyce Pereira[2]

[1]*Department of Chemistry & Biology, Federal Technological University of Paraná, Av. 7 de Setembro, nº 3156, CEP 80230-901, Campus Curitiba - Paraná - Brazil.* [2]*University Center FAE, St. 24 de Maio, nº 135, CEP 80230-080, Campus Centro - Curitiba -Paraná - Brazil.E-mail: cxavier@utfpr.edu.br.*

The pulp and paper industry in Brazil is in constant development and growth. Nationally, the annual pulp production should increase by 57% over the next year. Besides the paper production also should increase about 34%. Thus The process of making paper and cellulose generates large amounts of effluents, which mostly have lignins compounds, organic matter, color and toxicity. The aim of this paper was to analyze the mainly qualitative and quantitative aspects regarding to pulp and paper production in Brazil. The qualitative aspects related to the effluent were based on the determination of organic matter (BOD5 and COD), lignins, aromatics and phenolics compounds. Apart of it the biomass was also analyzed by solids (VSS and TSS) and total organic carbon (TOC) measurements. The samples of effluents and biomass came from two factories in the region of Curitiba. In the quantitative analysis it was emphasized the water consumption and effluent generation in Brazilian pulp and paper industries.

The effluents had different amounts of lignin, aromatic and phenolic compounds depending on the type of wood. However, the contents of organic matter like BOD5 of both industries were similar. The data from water consumption and effluent generation on Brazilian pulp and paper industries allowed to observe that the water consumption on this sector ranges from 25 m3.t-1 and 236 m3.t-1. The Brazilian pulp and paper mill presents an extended range on wastewater consumption and effluent generation. While some of them have European standard of best practices, others have the possibility to improve.

Keywords: Qualitative aspects; Quantitative aspects; Water consumption; Effluent; Pulp and paper mill

Water recovery from eucalyptus kraft bleaching effluents using combined UF/RO membranes and electrooxidation

C. Salazar, E. Soto, A. Vidal, H. Mansilla, R. Borquez, M.. Mondaca, C. Peña and C. Zaror

Environmental Science Center EULA-Chile, University of Concepción, Concepción, Chile.
E-mail: clsalazarn@udec.cl

Most cellulose production in Chile takes place in the BioBio region, where a large fraction of pine and eucalyptus plantations are located. Unfortunately, steady reductions in rainfall due to climate change have seriously affected water availability, threatening future industrial developments. In this context, efforts are being made to increase water circuit closure and reduce fresh water consumption in cellulose mills. On the other hand, membrane technology has experienced significant advances in recent years and has been identified as an interesting option for water recovery.
This study presents experimental results on the combined use of membrane separation technologies and electrochemical oxidation to recover water from ECF bleaching effluents in kraft cellulose production. Bleaching effluents from the first stage of ClO2 acid bleaching (D0) and alkaline extraction stage (E1), were collected from a hardwood kraft pulp mill located at the Biobío Region, in Central Chile. Effluents were stored in polyethylene containers of 20L and kept refrigerated at temperature below 4°C. Before the membrane filtration, cellulose fibres present in bleaching effluents were removed by microfiltration using popypropylene cartridges (5 and 1 um). Ultrafiltration (UF) was carried out using an Alfa Laval Mini-Lab 10 DDS (Naksov, Dinamarca) module, with a total area equal to 0.034m2 and two permeate outlets. Polysulphone Alfa Laval membranes with nominal cut-off sizes (MWCO) of 50, 20 and 5kDa were used. A potentiostat, Model 1100 A, CH Instrument; was used in electrochemical experiments. A pyrex temperature controlled electrochemical reactor (0.25L) was equipped with DSA anode, stainless steel 316 cathode and Ag/AgCl as reference electrode. Experiments were conducted at 2-4 volts and current densities in the range 30-250 A/m2 for up to 60 min. Raw acid bleaching effluents (pH 2.6) featured COD concentrations around 2 gO2/L, 0.8 g chloride/L and total phenols concentration 0.04 g/L; whereas alkaline effluents (pH 12.8) presented a COD of 1.8 gO2/L, 0.4 g chloride/L, and total phenols 0.01 g/L. Membrane permeabilities using distilled water were estimated around 120, 52, 80, and 25 (L/hm2bar), for 50, 25, 20, and 5 kDa, respectively, whereas RO permeability was less than 4 (L/hm2bar). UF membrane 50kDa was very effective to remove COD (67%), phenols (51%), and color (90%) from alkaline effluents. However, this membrane showed a poor performance when acid effluents were used. Indeed, only 14% of DQO, 35% phenols and 28% color were removed by 50kDa. Similar removal efficiency was found when other UF sizes were tried. On the other hand, reverse osmosis of UF permeate yielded COD, phenols, color and chloride removal around 98, 93, 98, and 97%, respectively. Moreover, conductivity was reduced by over 97% showing that most sodium, carbonate and other ions were also removed by RO. The high quality UF-RO treated permeate was tested as water source in D bleaching of eucalyptus predelignified pulp, showing no measurable adverse effect on pulp optical quality. Electrooxidation of retentate effluents from 20kDa UF led to 24% and 40% COD reductions for acid and alkaline effluents, respectively. The higher COD removal shown in the case of alkaline retentate effluents could be directly could be explained by the greater hydroxyl radicals generation rate due to the higher concentration of OH- ions. Hydroxyl radicals are known to act as oxidizing agents of organic compounds in indirect anodic electrochemical reactions (Soloman et al., 2009; Basha et al., 2010). It is concluded that combination of UF-RO allows significant removal of key contaminants from alkaline bleaching effluents, yielding a high quality permeate suitable for process recycle.

Acknowledgments: The authors would like to thank Project Basal PFB-27 and Fondecyt 1100738 for their financial support.

OPERATION OF TREATMENT PROCESSES TO MINIMIZE DISCHARGE AND NUTRIENT

Dynamics in the aeration basin of the activated sludge process within ECF kraft pulp mills

S. Toivakainen, T. Laukkanen and O. Dahl

Aalto University, School of Science and Technology, Department of Forest Products Technology, Clean Technologies Group, Helsinki, Finland, Tel +358 50 911 8746, sakari.toivakainen@aalto.fi

Concentration profiles of different wastewater qualities with respect to time were measured in plug flow aeration basins. Three activated sludge plants within ECF kraft pulp mills were studied, two of which were integrated pulp and paper mills. Concentrations at different stages of aeration were also studied in two pilot-scale activated sludge plants. The aeration time in the examined activated sludge processes was between 14-20 hours and sludge retention time varied from 17 to 30 days. In addition to wastewater concentration profiles, dissolved oxygen, oxygen uptake rate, pH and temperature profiles were also measured on site.

Most of the dissolved chemical oxygen demand (COD), biochemical oxygen demand (BOD) and adsorbable organic halogen (AOX) decrease took place almost immediately when influent and return sludge was mixed due to biosorption. Also, oxygen and nutrient uptake was high at the beginning of aeration and decreased gradually in a few hours after which only minor changes in concentrations or oxygen uptake occurred. The concentrations of total dissolved nitrogen, phosphorus and AOX can even rise when aeration time is extended probably due to the mineralization of biomass. As a conclusion, the aeration capacities in the activated sludge plants studied here allow modifications like sequencing aeration for chlorate reduction or biological phosphorus removal.

Keywords: aeration basin, activated sludge process, concentration profile, pulp mill effluent

Decolorization of Kraft pulp ECF bleaching effluent using the photocatalytic system UV / k2S2O8 : Optimization by experimental design.

M. Yeber and J. Novoa

Universidad Católica de la Santísima Concepción, Facultad de Ciencias. Concepción-Chile, mcyeber@ucsc.cl

Pulp industry produces high volumes of wastewater with non-biodegradable organic compounds and resistant to physicochemical treatments. Advanced oxidation processes show a potential for destruction of pollutants, improving the color and bioavailability of the effluent. Advanced oxidation processes (AOP's) generate in situ active radicals with high oxidation potential, allowing for the destruction of polluting agents through organic compound mineralization.

The use of experimental design is a tool that allows adjustment of factors to obtain an optimal response in experimental analysis. We used multivariate analysis to optimize the process of organic removal from wastewater in a cylindrical reactor, using UV radiation (254 nm) and peroxodisulfate, which is a strong oxidant generated in situ by irradiation sulfate radicals specie highly reactive (2.5-3.1 eV).

We used a factorial 2n design to obtain the best experimental conditions to efficiently remove the compounds from the solution, using pH, time in minutes and oxidant concentration in ppm as experimental variables. In order to verify the model, the reaction kinetics was carried out using the optimal values of the variables showing that the model is correct reaching the color removal in 99%. At the same time the TOC reduction reached 90%. The advanced oxidation process using K2S2O8/ UV system was efficient in the mineralization of the effluent by the action of sulfate radical. The chloride ions in solution could promote faster compound reduction due to the presence of sulfate radicals that oxidize chloride ions to chlorine radicals.

Keywords: Sulfate radicals, chlorine radicals, pulp and paper effluent, experimental design

Ability of Pseudomona aeruginosa bacteria to degrade Kraft lignin

M. Yeber and T. Silva

Universidad Católica de la Santísima Concepción, Facultad de Ciencias. mcyeber@ucsc.cl

The wastewater with high color generated during the production of bleached pulp process cause inhibition of biological activity when this effluents are released into the environment. *Pseudomona spp.*, is a gram negative and aerobic bacteria, that presents 1.5-5 µm in length and a diameter of 0.5 - 1.0 µm. It has polar flagella allowing their mobility in response to stimuli (chemotaxis) and also finds substrates in low concentrations. They are organisms ecologically important in the soil and water, and are probably responsible for the aerobic degradation of many soluble compounds derived from animal and vegetable organic matter in decomposition.

The aim of this study was to evaluate the capacity of the bacterium *Pseudomona aeruginosa* to degrade Kraft lignin as unique carbon source and energy for bacterial growth. Pure water soluble Lignin with molecular weight between 5000 - 28000 g mol -1 was used to prepare the solutions. Lignin samples were incubated with *Pseudomona aeruginosa* at 30° C for 5 days. Then the samples were analyzed measuring the absorbance at 460 nm using an UV-1601 SHIMADZU spectrophotometer, the viable cell were also counted to determine bacterial growth. A flow reactor was used to determine the degradation of Kraft lignin by the *Pseudomona aeruginosa* bacteria in biofilm. A bacterium was biofilms forming on polyethylene layer inside the bioreactor. The results show that with treatment at pH 8 the bacteria in biofilm removed the lignin with best efficiency, achieving 98% removal during the first day of treatment.

Keywords: Bacterial, *Pseudomona aeruginosa*, Kraft Lignin, Pulp and Paper Industry

Performance assessment of membrane separation technologies to recover water and chemical resources from segregated ECF bleaching effluents in Pinus radiata kraft pulp production

E. Oñate[1] , C. Salazar[2] and C. Zaror[1]

[1]*Faculty of Engineering, Department of Chemical Engineering, University of Concepción P.O Box 160-C. Concepción-Chile. E-mail: elizabethonate@udec.cl.*
[2]*Environmental Science Center EULA – Chile, University of Concepción. P.O. Box 160- C. Concepción-Chile.*

The aim of this study was to assess the performance of membrane separation to recover water and chemical resources from segregated bleaching effluents in cellulose production. Acid and alkaline effluents from Pinus radiata kraft pulp bleaching were obtained from local mills, and subjected to ultrafiltration (UF), nanofiltration (NF) and Reverse osmosis (RO) using a membrane filtration system Alfa Laval LabStak ® M20 with a range of 0.036-0.72 m2 of membrane area. UF cutoffs around 1, 2, 5, 10 and 20 KDa were used, whereas NF featured retention $\geq$ 98% $MgSO_4$ and 90-98% NaCl for RO. Experiments were conducted in «concentration mode» and «recirculation mode». Fouling, and solid-liquid interaction, such as z potential and contact angles were determined. Membranes surface morphology was determined by scanning electron spectroscopy. In the case of UF and NF processes complete blockage, standard, intermediate, cake filtration models, and combined models at constant pressure and flow were determined. The Chen fouling potential model was used to describe RO performance. All permeate and retentate effluents were characterized on the basis of physicochemical and toxicological parameters. Results show that alkaline and acid lines present quite different responses to membrane separation. The application of membrane separation processes as part of a water and chemicals recovery strategies is discussed under the light of experimental findings obtained here. It is concluded that such technologies could be applied providing that membrane physical-chemical properties are carefully selected according to the type of effluent to be treated.

Keywords: membranes separation technologies, performance assessment, segregated bleaching effluents, Pinus radiata kraft pulp, recover water and chemical resources

Evaluation of the bioaugmentation effect by a commercial-biological additives product for enhancing the performance of the biological paper mill wastewater treatment plant

M. Aqueveque[1], M. Cariñanco[2], G. Rivera[2], G. Pozo[3] and G. Vidal[3]

[1]*Universidad San Sebastián. Faculty of Engineering and Technology. Civil Biotechnology Engineering School. General Cruz #1577. Concepción-Chile. E-mail: maqueveqeg@gmail.com.*

[2]*Norske Skog BIO BIO. Continuous Improvement Department. Pedro Aguirre Cerda #1054, San Pedro de la Paz. Concepción-Chile.*

[3]*Environmental Science Center EULA – Chile, Universidad de Concepción. P.O. Box 160-C. Concepción-Chile.*

Bioaugmentation is the application of specifically selected bacteria into a wastewater treatment system to enhance the organic matter removal (Biological Organic Demand, BOD5 or Chemical Organic Demand, COD), for example. These commercial products are used in the wastewater treatment plant when a high BOD5 loading or a shock comes to the wastewater treatment plant. Also, it biological additive is used when biomass concentration in the system are not enough to reduce the BOD5 at the discharge level.

The goal of this research is to evaluate the performance of the commercial bioagumentation product to enhance the COD removal contained in a paper mill effluent treated by a biological system. The assays were performed under batch conditions, applying two experimental designs: 1) It was evaluated the application doses of the biological additive considering different fibre concentrations into the system; 2) the optimization of the doses application regarding the hydraulic retention time (HRT). In both cases an experimental design was done.

Results from the first experiment show that the biological additive do not improved the COD removal ($p>0.05$), even after 22 hours of HRT. Regarding the main effect, the experimental design show that there are not any significant difference between the performance of the reactors without (76.8%) and with fibres (72.7%). On the other hand, the BOD5 removal after 22 hours in all the assays was up to 93%. Moreover, the response surface design ($R2 = 97.9\%$) showed that the HRT is the principal factor involved in the performance of the COD removal ($p<0.05$). Furthermore, it was not find any correlation between biological doses and BOD5 removal efficiency. Finally, in this work the use of biological additive as a bioaugmentation strategy to improve the organic matter contained into paper mill effluent was not obtained.

Keywords: Bioaugmentation, paper mill effluent, COD, biological process

Pinus radiata and Eucalyptus globulus wastewater treatment from sawmill industry by constructed wetlands

C. Muñoz, I. Vera, C. Plaza de los Reyes and G. Vidal

Environmental Science Center EULA – Chile, University of Concepción. P.O. Box 160- C. Concepción-Chile.

Sawmill industry produces wastewater because of sprinkler irrigation on the log yards. For small producers, constructed wetlands (CW) can provide a low cost alternative for removing compounds contained into wood leachate. The goal of this work was to evaluate the treatment capacity of the constructed wetlands (CW) for removing organic matter and specific compounds coming from *Pinus radiata (PR)* and *Eucalyptus globulus (EG)* wastewater from sawmill industry.

In laboratory, mesocosm with and without *Phragmites australis* were used. The systems were fed through a system fed-batch with PR and EG wastewater, to increases of organic load rate (OLR) at the same time of hydraulic retention time of 7 days during an operation period of 98 days.

The results showed similarity in the behaviour of CW fed with PR and EG wastewater, finding greater removal efficiency in CW from 65% to 70% of COD, from 80% to 85% of BOD5, from 60% to 65% phenolic compounds and 30% to 70% of specific compounds in PR and EG, respectively. However, CW reached maximum removal efficiency at intermediate OLR (13 g COD m^{-2} d^{-1} in PR and 26 g COD m^{-2} d^{-1} in EG). The yield associated with the toxicity of the wood extractives compounds was reduced at the maximum OLR (19 g COD m^{-2} d^{-1} in *Pinus radiata* and 40 g COD m^{-2} d^{-1} in *Eucalyptus globulus*). This would be related to performance of CW in the removal of biologically active compounds present in effluents from sawmill.

CW would be alternative to removal organic matter (>65% of COD), specific compounds (>30%) and phenolic compounds (>60%) from PR and EG wastewater from sawmill industry as a technological solution to wastewater treatment for small producers.

Keywords: Sawmill Industry, *Pinus radiata*, *Eucalyptus globulus*, Constructed wetlands

Statistical analysis on the performance of a moving bed bioreactor fed with Eucalyptus globulus kraft mill effluent

C. Villamar[1], P. Aguayo[1], G. Pozo[1], K. Sáez[2], and G. Vidal[1]

[1]*Environmental Science Center EULA-Chile. Universidad of Concepción. Box 160-C, Concepción, Chile. E-mail: glvidal@udec.cl.*

[2] *Department of Statistics, Faculty of Physical and Mathematics Sciences, University of Concepción, Concepción, Chile*

The kraft mill effluents are characterized for organic matter content, color and toxicity. The MBBR systems (Moving Bed Biofilm Reactor) are a technological alternative for effluents with lower nutrients content in front of to conventional systems (activated sludge and aerated lagoon) and this permit also that can operates with more loading organic.

In this work was evaluated the effect of HRT (Hydraulic Retention Time) and nutrients relationship (BOD_5:N:P) in the organic matter, color, phenols totals removal and biomass (suspended and biofilm) evolution.

The MBBR was operated during 148 days in 4 phases using kraft Eucalyptus globulus effluents. The I and II phase were worked with a BOD_5:N:P of 100:3:1 and a HRT of 40 to 17 h, respectively. The III and IV phases were operated with a BOD_5:N:P relationship of 100:1:1 and a HRT of 8 to 4 h, respectively.

The results showed a maximum Biological Oxygen Demand (BOD_5) and Chemical Oxygen Demand (COD) removal of 98% and 69% respectively, color (-1 to 47%) and totals phenols (4 to 38%). The organic matter (BOD5: $p < 0{,}0001$, COD: $p = 0{,}0014$), color ($p < 0{,}0001$) and phenols totals ($p < 0{,}0001$) were significantly influenced for nutrient ratio. The suspended biomass was not affected significantly ($p = 0{,}1810$) by the nutrient ratio and showed better settleability conditions in all the operation. However, the biofilm varied significantly ($p = 0{,}0286$) by the nutrient ratio. During the operation, the biofilm was more than suspended biomass (> 75%).

Keywords: Biofilm, Eucalyptus effluents, nutrients ratio, MBBR

Optimization of culture conditions for lignin degradation by Anthracophyllum discolor in an airlift bioreactor

O. Rubilar[1]*, G. Tortella[1], K. Romero[1], C. Bornhardt[1,2] and M. Diez[1,2]

*[1]Scientific and Technological Bioresource Nucleus, Universidad de La Frontera, PO Box 54-D, Temuco, Chile. [2]Chemical Engineering Department, Universidad de La Frontera, PO Box 54-D, Temuco, Chile. *Corresponding author: Tel: +56 45 325487; fax: +56 45 325053, E-mail address: orubilar@ufro.cl.*

In Chile, millions of tons per year of pulp and paper are produced, generating large quantities of toxic and intensely colored waste effluents, causing severe water pollution. The primary contributors to the color and toxicity are high molecular weight lignin and its derivatives. The color is responsible for problems such as reduction in the light penetration depth of the receiving waters, which may reduce photosynthetic rates and affect aquatic productivity. A number of treatment processes have been suggested to enhance color removal, such as ultrafiltration and flocculation. However, to date no economically feasible solution has been established at industrial scale. A biological process that has attracted substantial interest is the use of white rot fungi, which may remove or reduce a wide variety of color generating compounds. In previous studies a strain of Anthracophyllum discolor was isolated from templated forests of southern Chile, which had been few reported in literature, and which demonstrated high ligninolytic activity and pollutant degradation potential. However, the optimal culture conditions for the production of enzyme have not been determined.

In this context, the potential of A. discolor for lignin biodegradation in an airlift reactor was evaluated. In the first stage of this study, different growths conditions of A. discolor were studied. The effect of pH, the addition of MnSO4 and Tween 80 were evaluated on ligninolityc enzyme production in shaken Erlenmeyer flasks. In the second stage, the degradation of different lignin concentrations by A. discolor in shaken Erlenmeyer flasks was determined. Finally, the lignin degradation process by A. discolor was scaled up in a 2.4 l airlift bioreactor containing 1000 mg l-1 lignin. Only the manganese peroxidase (MnP) activity was detected in the culture mediums evaluated. A continuous production was expressed in medium at pH 4.5 with a maximum of 68 ± 8 U l-1. When the culture medium was supplemented with MnSO4 (250 μM) and Tween 80 (0.05%), the maximum MnP activity increased to 470 ± 31 U l-1. In shaken Erlenmeyer flasks, the highest lignin degradation was obtained in the optimized medium (at pH 4.5, supplemented with 250 μM MnSO4 and 0.05% Tween 80) for maximizing MnP production. In the airlift bioreactor, 49.9% lignin and 61.0% color were removed. Comparing the MnP activity levels in the lignin degradation processes, a low MnP activity was detected in the airlift bioreactor in comparison with the MnP activity level detected in Erlenmeyer flasks, probably due to agitation stress. The results of this study make it possible to approximate the best culture conditions of A. discolor for lignin degradation.

Acknowledgments: This research was financed by the National Fund for Science and Technology (FONDECYT), Project 3080013 and partially by FONDECYT Project 11090678.

Application of a predictive model of pulp and paper mill effluent treatment plant control parameters by means of GOLDSIM model

F. Canales, F. Dibarrart and C. Muñoz

AMEC, Environment & Infrastructure, Santiago-Chile. E-mail: franklin.canales@amec.com.

One of the most outstanding areas of the forestry industry in Chile, in terms of projection, production and foreign currency, is related to pulp and paper production. The development of this economical activity has shown an ongoing technology modernization in most of Chilean pulp mill effluent treatment plants. These new technologies have enhanced both primary and secondary treatments (ranging from aerated lagoons to activated sludge) and in certain cases, tertiary treatment as well. This work has two main objectives: firstly, to carry out a review of the effluent treatment systems used in the Chilean mills, and secondly, to model a typical effluent treatment plant based on these technologies. The model could be used to predict the steady and dynamic behaviour of the plant. To accomplish the latter, a GOLDSIM simulation model has been developed in order to estimate and predict the behavior of the main system parameters. This software could simulate time depending systems as well as static ones. Thus, the intention is to develop a model that can be used in different industrial-scale cases as an operation controlling and effluent quality predicting tool.

Keywords: pulp and paper mill, effluent treatment plants, simulation

RESOURCE AND ENERGY RECOVERY OR CONSERVATION

Polyhydroxyalkanoates (PHA) biosynthesis from kraft mill effluents in a moving bed bioreactor (MBBR): operational factors, biomass origin and nutrients effect

G. Pozo[1], C.Villamar[1], M. Martínez[2] and G.Vidal[1]

[1]*Environmental Science Center EULA – Chile, University of Concepción. P.O. Box 160- C. Concepción-Chile. E-mail: gpozo@udec.cl.* [2]*Microbiology Department, University of Concepción. P.O. Box 160-C, Concepción, Chile.*

In the MBBR operation, nutrient diffusion into the biofilm layer is highly limited, which the bacterial populations adopt specific survival strategies. In particular, cells often accumulate organic carbon as internal polymers, such as polyhydroxyalcanoate (PHA). In this work the operational factors and nutrient effect on the biosynthesis of PHA were studied using kraft effluents and MBBR technology, as well as three kind of sludge origin was studied using batch system. The MBBR was operated during 225 days in five phases. The Hydraulic Retention Time (HRT) was reduced from 48 to 4.8 h between phases I and IV, the BOD_5:N:P ratio (100:5:1 and 100:1:0.2) was evaluated as an operation strategy in phases IV to V.

Inocula were obtained from activated sludge plants from sewage (SAS), paper mill (PAS) and kraft (KAS) for batch assays. The maximum PHA accumulation was obtained under a BOD_5:N:P relationship of 100:1:0.2 using KAS sludge with 30.40% of cells. Yields obtained ranged from 0.10–0.14 mg PHA/mgCOD. In the MBBR operation the maximum absolute fluorescence against the proportion of cells accumulating PHA was obtained for a OLR of 1.19 ± 0.04 $kgBOD_5/m^3{\cdot}d$ and a BOD_5:N:P relationship of 100:1:0.2. The increase of PHA biosynthesis is due to the OLR increase and it is not attributable to the increase of cell concentration, which is maintained constant in stationary status during bioreactor operation. However the sludge performance when the HRT was reduced from 48 to 4.8 h shown that the fraction of suspended biomass decreased (33.4%) and biofilm growth in the MBBR increased to 84%.

Keywords: Kraft mill effluent, MBBR, sludge, polyhydroxyalkanoates.

Photovoltaic assisted electrocoagulation of a paperboard manufacture plant effluent using aluminum electrode and ash leachate

M. Mwamburi[1], M. Mutiso[2], O. Orori[2]., K. Senelwa[2], K. Balozi[2] and L. Etiégni[2]

[1]Department of Physics, Chepkoilel University College, P.O. Box 1125-30100 Eldoret, Kenya, mghendi@yahoo.com.
[2]Department of Forestry & Wood Science, Chepkoilel University College, P.O. Box 1125 Eldoret.
Email:miry.mtiso@yahoo.com, benorori@yahoo.com, ksenelwas@yahoo.co.uk, balozibk@hotmail.com, lazetiegni@amatala.org

The treatment of wastewater from pulp and paper mill effluent by electrocoagulation can be an energy intensive undertaking making the process uneconomical. In this paper we evaluate the possible use of a commercial photovoltaic (PV) unit to drive electrochemical reaction in an electrocoagulation cell for the treatment of a paperboard mill effluent. The electrodes were made of aluminum. The raw effluent volume in the electrocoagulation cell was kept constant except for the wood ash leachate which varied from 10 to 40 ml. The electrocoagulation cell was fitted with a photo meter, temperature sensor, current and voltage monitoring devices. The results indicate that the I-V characteristics of the PV unit auto-regulated the electrode potential in line with the conductivity of the effluent. Higher wood ash leachate concentrations in the mill effluent increased wastewater conductivity, which accelerated the decolourization reaction and shortened the rise in the treated effluent temperature. However optimum PV operation varied with solar irradiation intensity. Over time, when the coating on the electrodes was removed, there was a sharp reduction in the electrocoagulation cell voltage and light transparency but with an increase in the cell current. Effluent parameters (BOD, COD, TSS) were substantially reduced by between 70 and 95%. This study shows that PV cells can directly power an electrocoagulation cell without a battery or charge regulator. PV cells can reduce electrocoagulation electric energy at minimum capital cost and minor operation and maintenance costs because the absence of moving parts allows for long life (20-25 years).

Keywords: aluminum; electro-coagulation; photovoltaic cell; pulp and paper; wastewater

Conversion of an aerobic treatment plant of pull mill effluent to an anaerobic process saving energy and nutrients, with sludge production reduction and bioenergy production

S. Montalvo[1], P. Madriaga1 and L. Guerrero[2]

[1]*Environmental Biotechnology Laboratory, Chemical Engineering Department, University of Santiago de Chile, Ave. Lib. Bdo. O'Higgins 3363, Estación Central, Santiago de Chile. Email: silvio.montalvo@usach.cl.* [2]*Department of Chemical and Environmental Engineering, Federico Santa María Technical University, Ave. España 1168, Valparaíso, Chile. Email: lorna.guerrero@usm.cl*

The main objective of this work is technically and economically evaluate the conversion of an aerobic process in anaerobic process, saving and energy recovery in a treatment plant of liquid effluent from a pulp and paper industry.
The work was carried out after the completion of the following activities: a) determining the cost efficiency and current associates of aerobic processes installed in the wastewater treatment plant, b) development of a technological proposal for conversion from aerobic to anaerobic c) assessment of economic aspects related to the proposed technological change; d) technical and economic comparison of the proposed amendment to the existing technology.

From the results we conclude that it is technically and economically feasible, the conversion of aerobic process in an anaerobic process with energy savings and achieving other economic and environmental benefits.
Comparing the current process with the proposed amendment shall include the following benefits, among others:

- Reduction in energy consumption, estimated at 906.5 [kwh].
- Decrease in nutritional requirements, estimated at 17,535 [kg / month] of granulated urea and 7099 [kg / month] of phosphoric acid.
- Decrease in the volume of sludge generated in the secondary treatment at 153 [ton / month], which are more stabilized than in aerobic processes.
- CO^2 emission reduction estimated in 5735 [tonCO^2/año].
- Biogas generation of high fuel value, measured at an average flow of 8308 [m^3/d].

In the economic comparison between the current process and the proposed amendment, it was demonstrated that the most significant savings are due to the decrease in electricity, estimated at U.S. $ 348,276 per year, this is followed by savings from reduced nutrient U.S. $ 153,696, while the savings from decreased production of sludge is U.S. $ 25,080 per year
Acknowledgements. This work was supported by DICYT PROJECT 091111MM, University of Santiago de Chile.

Keywords: aerobic and anaerobic treatment, plant conversion, pulp mill effluent

Production of polyhydroxyalkanoates (PHA) by moving bed biofilm reactor (MBBR) treating a paper mill wastewater

M. Jarpa[1], R. Baeza[1], G. Pozo[1], M. Martinez[2] and G. Vidal[1]

[1]*Environmental Science Center EULA – Chile, University of Concepción. P.O. Box 160- C. Concepción-Chile.*
E-mail: mayjarpa@udec.cl. [2]*Microbiology Department, University of Concepción, P.O Box 160-C. Concepción-Chile.*

The effluent of the paper industry contains organic matter that have to remove before to be discharged. Because of this, a potential by-product production like polyhydroxyalkanoates (PHA) could be a valorization alternative for this wastewater. The goal of this work was to evaluate BOD_5:N:P relationship in the PHA production using the wastewater generated in this type of industry. The PHA production process was developed under MBBR technology which was continuously operated (300 d). In the first step, the wastewater was fed under BOD_5:N:P relationship 100:5:1 and after to (270 days) decreased to 100:1:0.3. The system was operated with an organic load rate (OLR) which was increased from 0.20 to 7.67 kg COD/ m^3•d. The concentration of acetate, propionate and butyrate, the main volatile fatty acids (VFAs) was 0.025 g/L, 0.034 g/L and 0.014 g/L. The maximum BOD_5 removal was 98.7 %, for relationship of the 100:5:1. Meanwhile, for BOD_5:N:P relationship 100: 1:0.3 was between 83.9 - 87.2 %. The maximum PHAs accumulation was obtained when the BOD_5:N:P relationship was 100:5:1. Under this conditions, the fluorescence intensity related with PHA accumulation was ranged between 17.40 - 173.40. A higher PHA accumulation rate was obtained when the OLR was 7.67 kg COD/ m^3•d, 85.4 % of total cells measured through flow citometry. On the other hand, for the BOD_5:N:P relationship of the 100:1:0.3 the fluorescence was between 27.05 - 59.62.

Acknowledgements: Fondecyt Grant No. 1070509, CONICYT grant (24100082), Papeles Norske Skog Bio-Bio Ltda.

Keywords: Nutrients, Paper mill wastewater, MBBR, PHA, Bioplastic

FATE AND EFFECT OF SPECIFIC COMPOUNDS DURING WASTEWATER TREATMENT

Degradation of 2, 3, 4, 6 -Tetrachlorophenol by Pseudomonas sp. isolated from the sludge of pulp and paper industry

S. Kumar[1], S. Chakrabarti[2] and M. Sudhakara[1].

[1]*Department of Biotechnology & Env. Sciences Thapar University, Patiala – 147004, Punjab, India.*
[2]*Thapar Centre for Industrial Research and Development, Paper Mill Campus, Yamunanagar – 135001, Haryana, India;*
E mail: swapan@thaparpaperresearch.org, Phone: +911732292703, FAX: +911732292748

2,3,4,6-tetrachlorophenol (2,3,4,6-TeCP) finds widespread industrial use as biocide in wood treatment, flame retardant, preservative for leather and textile goods, solvent and reagents in synthetic chemistry. The compound is reported to be toxic, and persistent in soils, sediments, and natural water as well as in food and human urine. Wastewater and sludge in pulp and paper mills contain the chlorophenol which arises from the bleaching process of kraft pulp.Very few bacterial isolates so far are reported for the degradation of 2,3,4,6-TeCP. In the present study a bacterium (CL7) were isolated from the sludge by enrichment technique and analyzed for 2,3,4,6-TeCP degradation. The isolate showed higher degradation efficiency, 85% degradation at 200 mg/l. With increased concentration to 600 mg/l the degradation efficiency decreased to 73%. The isolate was morphologically, biochemically and molecularly (based on 16s rRNA gene analysis) characterized. 16s rRNA gene sequence indicated maximum similarity with *Pseudomonas* sp. The optimum pH and temperature for the degradation of 2,3,4,6-TeCP were observed at 7.5 - 8.5 and 370C respectively. The study indicates that the isolate has better potential to degrade 2,3,4,6-TeCP at high concentration.

Keywords: Tetrachlorophenol, sludge, degradation, pulp and paper mills, Pseudomonas sp

Removal of recalcitrant organic compounds in pulp and paper effluents under advanced oxidation processes (AOPs): a review

M. Jarpa, C. Baeza and G. Vidal.

Environmental Science Center EULA – Chile, University of Concepción. P.O. Box 160- C. Concepción-Chile.
E-mail: mayjarpa@udec.cl.

Toxicity attributed to recalcitrant organic compounds in pulp and paper industries comes from wood extractives compounds, absorbable organic halogens (AOX), tannins, lignin and lignin derivatives. Extractives are lipophilic compounds (log K_{OW}> 4) released in the productive process. Resin acids and phytosterols, for their stability and resistance in structure had been found in effluent previously treated. Typical concentrations for resin acids ranged from 0.008 to 1.36 mg/L and in the case of phytosterols the followings values had been found: 165 μg/L for β-Sitosterol, 61.2 μg/L for Campesterol and 21.1 μg/L for Stigmasterol. These trace compounds may cause deleterious environmental impacts upon direct discharge to receiving waters, causing effects in exposed fish, mainly as endocrine disruptors.

For the removal of these compounds, advanced oxidation processes arise (AOPs) as efficient technologies. AOPs systems are based on the generation of hydroxyl radicals (OH•), a powerful oxidizing agents well known for transforming a great variety of recalcitrant organic compounds. Studies in these effluents with Fenton reaction (H_2O_2/Fe^{2+}) and photo-Fenton reaction ($UV/H_2O_2/Fe^{2+}$), register removal up to 96.4% TOC (Total Organic Carbon), 93.0% AOX, 84.6% color and 100% phenolic compounds. For O_3/UV combined with conventional wastewater treatment, literature showed removals up to 81.2% color and 70.0% phenolic compounds, 59.1% TOC and 75.5% COD (Chemical Oxygen Demand) and 95.0% AOX. While for UV/TiO_2, previous studies obtained approximately 50.0% COD, 40.0% color, 95.0% AOX, 79.6% TOC and 94.0 % toxicity removals.

Keywords: pulp and paper industries, advanced oxidation process, recalcitrant organic compounds, removals efficiency

Acknowledgements: Fondecyt Grant No. 1070509, CONICYT Grant (24100082).

Antibiotics effects on anaerobic biomass from kraft pulp mill effluent

L.Fernandes, S. Sáez, R. de Paula and C. Xavier

Department of Chemistry & Biology, Technological Federal University of Paraná, Av. 7 de Setembro, nº 3156, CEP 80230-901, Campus Curitiba - Paraná - Brazil. E-mail: lucas.alberto91@yahoo.com.br

In this work, it was evaluated the effects caused by two commercial antibiotics – Amoxicillin and Bactrim (sulfamethoxazole and trimethoprim) – on: (i) methanogenic activity, (ii) degradation kinetics and (iii) characteristics of biomass from a treatment plant of kraft pulp mill effluent. The analysis of the specific methanogenic activity (SMA) was used to determinate the kinetic parameters of anaerobic degradation (Andrews´ kinetic model) performed by the biomass in the presence of each antibiotic. After that, the biomass was characterizated by optical microscopy.

In the presence of Amoxicillin, it was observed activation of biomass; however the presence of Bactrim caused inhibition of the same. In the absence of both antibiotics, the biomass activity was reduced, which indicates the occurrence of permanent effects on it. Regarding the kinetics, it was obtained non-zero values for maximum rate of degradation (3.56 × 10-2 gCOD. gVSS-1.d-1) and very low inhibition coefficient for the biomass in the presence of both antibiotics. The higher values of affinity constant for the antibiotics (4.45 × 103 mg.L-1 for Bactrim and 1.19 × 102 mg.L-1 for Amoxicillin) indicate a higher affinity of biomass for Amoxicillin. The microscopic analysis of biomass exposed to both antibiotics show different effects depending mainly on their action range, their concentration, the biomass early composition and other factors. The presence of xenobiotic substances changes microbiological characteristics and composition of the biomass; it could interfere on effective wastewater treatment, because the removal of susceptible organisms causes an imbalance in the community, affecting the activity and ability of biomass degradation.

Keywords: pulp mill effluent, anaerobic treatment, specific methanogenic activity, antibiotics, amoxicillin, bactrim

Removal of endocrine disruptors and pharmaceuticals compounds from water using forestry by-products as adsorbents

M. Antunes, S. Pinto and F. Braga

Department of Chemistry and CQVR, Universidade de Trás-os-Montes e Alto Douro, Apartado 1013, 5001-801, Vila Real, Portugal. E-mail:cguiomar@utad.pt.

During the last decades, research scientists and governmental authorities became increasingly concerned about human and wildlife exposure to emerging contaminants, even at very low levels. Despite the existence of well-established adsorption and ultrafiltration technologies used to eliminate endocrine disruptors and pharmaceuticals compounds from water, economic considerations have stimulated the accomplishment of alternative studies to identify or develop new sorbents with a better cost-efficiency relation.

In this work, the sorption capacities of pine bark and almond shells for the removal of bisphenol A, 17β-estradiol, caffeine and paracetamol from aqueous solutions were evaluated. Samples of these traditional agro-forestry by-products were milled, sieved into different particle size fractions (0.10-0.15 mm and 1.5-2.0 mm) and submitted to two different kinds of treatment. Sorption experiments were conducted in batch system at room temperature and at natural pH. The sorption equilibrium was attained after 48 h, for all systems under studied. All tested compounds were adsorbed on the smaller particles of the sorbents more readily and effectively. The differences observed among the adsorption capacities are discussed in terms of physico-chemical characterization of the materials, which revealed the presence of meso and macropores.

The sorption equilibrium data were analysed using Langmuir and Freundlich isotherm models, with general results indicating that utilization of both materials as alternative sorbents for contaminated waters is promising due to availability in large amounts, and acceptable cost-efficiency ratio when compared with traditional adsorbents.

Keywords: pine bark, almond shell, endocrine disruptors, pharmaceuticals compounds, sorption

Chronic effects of Pinus radiata and Eucalyptus globulus Kraft mill effluents and phytosterols on Daphnia magna

D. López[1], S. Chamorro[1], J. Silva[2], E. Bay-Schmith[2] and G. Vidal[1]

[1]*Environmental Science Center EULA – Chile, University of Concepción. P.O. Box 160- C. Concepción-Chile. E-mail: dlopez@udec.cl.* [2]*Ecotoxicology Laboratory, Faculty of Natural Science and Oceanography, University of Concepción, P.O. Box 160-C, Concepción, Chile.*

Chronic effect on Daphnia magna of the kraft mill effluent corresponding processes working with Pinus radiata and mixed (50% Pinus radiata and 50% Eucalyptus globulus) as raw material and phytosterols (β-sitosterol and Stigmasterol) were evaluated.

The percentage of allometric growth rate (% AGR) was evaluated as the relationship between body length and body width. Also, first brood average time, broods average and the total neonates average were evaluated during the 21 d of the assays (with special evaluation at 7, 14 and 21 d). Indicators were evaluated respect to the positive control (17α-ethynylestradiol: EE2) and a blank.

The pine KPME exerts a stronger estrogenic effect on D. magna than the effluent of mixed alone. An alteration in body proportions was as intense in both effluents as in the exposure to 17α-ethynylestradiol. Other molecules apart from β-sitosterol and Stigmasterol must be contributing to the mentioned allometry. The phytosterols per se are responsible for 12.9 and 8.1% of the deviation from the natural shape, while the KPMEs account for 25.6 to 27.8% of shape deviation. Reproductive parameters, on the other hand, were affected mainly by mixed KPME. The role of β-sitosterol and Stigmasterol is discussed in relation to endocrine disruption. Acknowledgements. This work was supported by Fondecyt 1070509.

Keywords: Wood pulp wastewater, chronic effect, endocrine disruption, *Daphnia magna*; *Eucalyptus globulus*, *Pinus radiate.*

Photodegradation of dye remazol brilliant blue-r by HSO5- catalysed by Fe 2+

L. Bascuñan and J. Fernandez

Catholic University of the Santísima Concepcion, Faculty of Sciences, Alonso de Ribera 2850, casilla 297 Concepción, Chile. e-mail: jfernandez@ucsc.cl

In this work, the azo dye Remazol Brilliant Blue-R [RBB-R] was abated by peroxymonosulfate catalyzed by ferrous ions (HSO_5-/Fe^{2+}) under visible light. Mineralization and toxicity of the treated solutions were also evaluated.

Experimental planning was used to evaluate the weight of the following factors on the degradation: [HSO_5-] (X_1), [Fe^{2+}] (X_2) and [RBB-R] (X_3) the chosen response factor (Y) was the residual color after 15 minutes irradiation obtaining the following polynomial expression:

Y = 3,84 – 12,38X1 – 4,78X2 + 2,92X3 + 6,84X12 + 3,17X22 + 3,76X1X2 - 2,36X1X3 - 1,06X2X3

Optimal conditions were found at [RBB-R] = 100 ppm, [HSO_5-] = 0,25 mM and [Fe^{2+}] = 1,5. In this condition a total color removal is obtained after 20 minutes, but only a 20 % of organic matter removal is obtained after 300 minutes of irradiation. The bacterial growing of pseudomona aureuginosa showed that after the treatment the bacteria grown faster with an UFC/mL of $5,11x10^{12}$, that shows that the treatment clearly diminished the toxicity of the solutions that allows to use this treatment as a previous step to biological treatments.

Keywords: Remazol Brilliant Blue-R (RBB-R), Oxone, Degradación fotocatalítica, Pseudomona *aeruginos*

Characteristic of the kraft mill effluents from Pinus radiata and Eucalyptus globulus by means combination of yeast bioassays and CG-MS chemical analysis

S. Chamorro[1], D. López[1], R. Baeza[1], A. Olivares[2], B. Piña[2] and G.Vidal[1]

[1]*Environmental Science Center EULA – Chile, University of Concepción. P.O. Box 160- C. Concepción-Chile. E-mail: schamorr@udec.cl.* [2]*Institute of Environmental Assessment and Water Research (IDAEA-CSIC). Jordi Girona, 18. 08034 Barcelona, Spain*

The goal of this work is to evaluate the characteristic physico-chemical of effluent of *Pinus radiata* and *Eucalyptus globulus* used as raw material. These effluents were subjected to biological treatment before their analysis. Solid phase extraction (SPE), are performed with solvents of increasing polarity to determine most of the compounds analyzed through gas chromatography – mass spectrometry (GC-MS). For determine activity of kraft pulp mill effluents (*P. radiata*, and *E. globulus*) was evaluated to the Yeast Estrogen Screen (YES) assay, while Ligands of Aryl Hydrocarbon Receptor (AhR) was evaluated by Recombinant Yeast Assay (RYA).

By SPE and GC-MS were detected the main groups present in pulp mills effluents, which correspond to fatty acids, hydrocarbons, phenols, sterols and triterpenes. Activity of YES in *E. globulus* and *P. radiata* effluents showed values ranked between 1.475 and 0.383 ng/L of EE2 eq. RYA showed an effective concentration of AhR ligands significantly higher *E. globulus* than in *P. radiata* effluents. Analysis of their respective composition by SPE and GC-MS, suggested that aromatic compounds were the main determinants activity in the samples. In particular, benzaldehyde derivatives appeared as responsible for the activity, due to their structural properties and their high concentration in AhR ligand-rich samples.
Acknowledgements: This work was supported by Fondecyt Grant No. 1070509 and CONICYT 24080124 Doctoral Grant.

Keywords: *Pinus radiata, Eucalyptus globulus,* SPE-extraction, GC-MS, yeast bioassays

Bioaugmentation with fungal pellet formulated with lignocellulosic waste to degrade liquid residue of atrazine in a biobed reactor

S. Elgueta[1], R. Cuevas[1], M. Campo[1], G. Cofre[1] and MC. Diez[2]

[1]*Scientific and Technological Bioresource Nucleus, Universidad de La Frontera, P.O. Box 54-D, Temuco, Chile. E-mail: selgueta@ufro.cl.* [2]*Chemical Engineering Department, Universidad de La Frontera, P.O. Box 54-D, Temuco, Chile.*

Unsatisfactory management of pesticides especially pesticide handling pose major risks for contamination can give rise to residues in surface waters and groundwater and in large volumes of soil. Biobed is a simple and effective alternative for the biotreatment of pesticides, these systems are easy to implement on-farm, are based primarily on ability to degrade pesticides through microbial communities especially white-rot fungi. Much attention has been directed toward the use of white-rot fungi for bioaugmentation to increase the biodegradation of pesticides mainly due to extracellular ligninolytic enzyme. One of the barriers to successful implementation of fungal bioaugmentation is the development of inexpensive and high quality fungal pellet. These inocula are in the form of pelleted solid supports. Therefore, the aim was to evaluate the effect of fungal pellets in a biobed contaminated with the atrazine. Fungal pellets of *Stereum hirsutum* were formulated with a mixture of different supports, based on lignocellulosic wastes, oligosaccharides and salt materials. The biobed was prepared by mixing an allophonic top soil (Andisol), commercial peat, wheat straw in a volumetric proportion of 1:1:2 and was inoculated with fungal pellets of *Stereum hirsutum* (10% w/w). The biobed was contaminated with 80 mg kg-1 of atrazine. After 30 days of incubation at 20ºC, ligninolytic enzyme activity, FDA, and respiratory activity were studied. The atrazine was measured by HPLC. The biodegradation of atrazine was higher for fungal pellets (55%) more than control (biobed non-inoculated with fungal pellets). At 30th day all the biological activities were higher for the biobed inoculated; the FDA activity (1.574 µg FDA/gr soil), the respiration (3054 ± 219 µg CO2 g-1) and total ligninolytic enzyme activity (1.87 ± 0.1 U kg-1) compared with control, FDA activity (1.053 µg FDA/gr soil), the respiration (1434.4 ± 113 µg CO2 g-1) and total ligninolytic enzyme activity (0.45 ± 0.1 U kg -1). In conclusion, the fungal pellets of *Stereum hirsutum* promotes better performance as support for fungal pellets to be used on biobed contaminated with atrazine. Acknowledgments: FONDECYT 1090678.

Keywords: Fungal pellet, Lignocellulosic waste, biobed

MANAGEMENT OF WASTE SOLIDS AND THEIR EVALUATION AS BY-PRODUCTS

Grits effect in physical and chemical aspects in co:composting with kraft mill secondary sludge

M. Zambrano[1,2], R. Cuevas[1,2] A. Rojas[2], M. Alvear[3], M. Villarroel[2] and G. Vidal

[1]Scientific and Technological Bioresource Nucleus, Universidad de La Frontera, P.O. Box 54-D,Temuco, Chile. E-mail: mzambra@ufro.cl. [2]Chemical Engineering Department, Universidad de La Frontera, P.O. Box 54-D, Temuco, Chile. [3]Chemical Sciences and Natural Resources Department, Universidad de La Frontera, P.O. Box 54-D, Temuco, Chile. [4]Environmental Science Center EULA – Chile, University of Concepción. P.O. Box 160- C. Concepción-Chile.

Segregation and alternative uses for the waste generation are the main focus for waste management. Wastes, like kraft mill secondary sludge, pine bark and grits obtained from the causticizing process are under evaluation for reuse. Secondary sludge usually coming from biological wastewater treatment, grits and pine bark have been characterized in terms of physico-chemical parameters. Composting is a more environmentally sustainable alternative than landfill for kraft mill waste disposal. Also its use as an organic fertilizer, soil amendment or supplemental nutrient source for potting media was demonstrated. From this point of view, organic compounds contained in sludge can have potentially beneficial effects on degraded soils (Total Organic Content, TOC of 37%). Moreover, the use of grits as inorganic amendment in composting can improve the process performance due to their higher alkalinity and buffer capacity. The purpose of this work is to determine the physical, chemical and hazardous waste and to determine the effect of different doses of grits in the physical and chemical properties during its co:composting with kraft mill secondary. Characteristics of dangerousness (inflamabiliad, reactivity, corrosivity) and toxicity (due to the presence of inorganic compounds) are lower than the ranges set by the EPA standards. In addition the chemical characteristics (pH (12.7), Ca (37.5), Mg (0.34%), P (0.30%) and K (0.11%) together with the micro-nutrients (0.47%) iron and zinc (0.12%) present in the sludge suggest its use as a promising alternative in composting waste. There was not generation of leachate in most of the tests, except in those with smaller dose of dregs (5 and 6%). Moreover, nutrients concentrations were not detected in the leaches generated. Acknowledgments: This research was financed by DIUFRO DI10-0010.

Keywords: Kraft mill solid wastes, grits, composting process

Biochemical parameters and toxicological characteristics of compost obtained from grits, biological sludge and bark coming from kraft mill solid wastes industry

M. Zambrano[1,2,] C.Torres[2], A. Rojas[2], H. Leal[2], M. Alvear[3], M. Villarroel[2] and G. Vidal[4]

[1]*Scientific and Technological Bioresource Nucleus, Universidad de La Frontera, P.O. Box 54-D, Temuco, Chile. E-mail: mzambra@ufro.cl.* [2]*Chemical Engineering Department, Universidad de La Frontera, P.O. Box 54-D, Temuco, Chile.* [3]*Chemical Sciences and Natural Resources Department, Universidad de La Frontera, P.O. Box 54 D, Temuco, Chile.* [4]*Environmental Science Center EULA – Chile, University of Concepción. P.O. Box 160- C. Concepción-Chile.*

Application of unstable compost can cause fitotoxicity inhibiting the growth of plants and could cause damages to the cultures that compete by oxygen or causing fitotoxicity to the plants due to the insufficient degradation of the organic matter (Brodie *et al.*, 1994). Growth of plants and tests of germination seeds are the techniques commonly used to evaluate maturity of compost. In this context quality of three compost with different grits addition was evaluated by means of stability and maturity evaluation. The compost obtained from kraft mill secondary sludge with bark and different content of grits 5, 6, 8, 10, 11% and humidity was assessed through biochemical assays (fluorescein diacetate, dehydrogenase, urease activities) to evaluate stability. In addition, germination index of *Raphanus sativum* and *Lycopersicon esculentum* L., toxicological test were used to evaluate maturity. Enzymatic activities for the compost evaluated presented similar values to those reported by other investigators of radish seed (*Raphanus sativum*) in compost of kraft pulp mill fulfills the norm of compost Chilean 2880 (80%). Two-Way ANOVA analysis determined that both seeds germination index and compost (independent variables) are directly related however, *Raphanus sativum* showed a significant effect ($p<0.05$) in relation to *Lycopersicon esculentum* L.. The ToxiChromopad test showed presence of fitotoxic substances only in compost with a high doses of grits (10% and 11%).

Acknowledgments: This research was financed by DIUFRO DI10-0010 and partially DIUFRO DI08-0011.

Keywords: Kraft mill solid wastes, grits, compost

Use of lignocellulosic wastes on biobed system to degrade pesticides

C. Urrutia[1], S. Fernández[2] and MC. Diez[2]

[1]*Graduate Program in Natural Resources Science, Universidad de La Frontera, Temuco*
Email: cynthia.urrutia.m@gmail.com. [2]*Department of Chemical Engineering, Universidad de La Frontera, Temuco*

To reduce the point sources contamination of pesticides include the biobed system which retains and degrades pesticides. The biomixture of biobed consists of straw, peat and topsoil (2:1:1 by volume). Straw guarantees a continuous supply of nutrients and high microbial activity, especially of lignin-degrading fungi, producing extracellular non-specific ligninolytic enzymes. These enzymes have been involved in the degradation of pesticides.

Great quantity of forest and agricultural wastes are produced annually in Chile. For example, the post harvest residues in 2009 of wheat, barley and oats were 592,723; 48,175 and 156,895 (ton year-1), respectively and the forest wastes were 1,078,121 (ton year-1).

In this study the effect of different lignocellulosic wastes to prepare a biomixture of a biobed system to treat pesticides were evaluated. The lignocellulosic wastes used were: barley husk (BH), sawdust (SD), oat husk (OH) and straw (ST) in different volumetric proportions with soil and peat. The traditional Swedish biomixture (peat:soil:straw, 25:25:50% v/v) was used as control. Pre-incubation time of the biomixtures, adsorption and degradation of pesticides (atrazine, isoproturon and chlorpyrifos) and microbial activity were analyzed for all biomixtures.

The best pre-incubation time for all biomixtures was between 30 and 40 days. The biomixtures with sawdust were the most efficient to adsorb pesticides, although biomixtures with barley and oat husk showed the highest biological activity (FDA, CO2 evolution and dehydrogenase) during degradation time. However, peroxidases were not detected in the biomixtures with barley husk probably due to high phenolic compounds content.

Acknowledgements: This study was financed by Fondecyt 1090678 and Innova-Corfo 09FCO2-6021 projects.

Keywords: Lignocellulosic wastes, biobed, pesticides

Characterization of biochars derived from agricultural and forestry residues

M. González[1], A. González[2], N. Sepúlveda[2], C. Toro[2], M. Cea[2], MC. Diez[2,3] and R. Navia[2,3]

[1]Doctoral Program in Sciences of Natural Resources, Universidad de La Frontera, P.O. Box 54-D, Temuco, Chile. (ma. eugenia.g.q@gmail.com). [2]Scientific and Technological Bioresource Nucleus, Universidad de La Frontera, P.O. Box 54-D, Temuco, Chile. [3]Department of Chemical Engineering, Universidad de La Frontera, P.O. Box 54-D, Temuco, Chile

A number of studies have suggested that terrestrial land application of biochar could effectively sequester carbon in soils and thus mitigate global warming. When biochar is applied to soils, it may also present other potential advantages including improvement of soil fertility and crop productivity, by increasing soil nutrients and water holding capacity.

However, the applications and functions of the biochar are highly depending on their physicochemical properties, therefore it is very important to characterize biochar before use.

In this study, biochar produced by pyrolysis of oat hull at 300 and 450 °C (BO300 and BO450) and pine bark at 300 and 500 °C (BP300 and BP500) were physicochemical characterized.

The result showed that the mineralogical, structural and chemical properties depend on raw material used and pyrolysis temperature. The patterns of X-ray diffraction (XRD) showed the presence of sylvite (KCl) in BO300 and BO450 samples. While XRD for BP300 and BP500 showed the presence of anorthite (SiO_8), calcite ($CaCO_3$) and quartz (SiO_2).

The specific surface area (BET) increased as the pyrolysis temperature increased. The BET area in the BO300 and BP300 were 0.1 and 1.9 m^2/g, respectively. While values at major temperatures the BET area were 6.6 and 175 m^2/g for BO450 and BP500, respectively.

The BP500 sample presented the highest content of the free bases. On the other hand, the BO450 sample presented a high pH (9.57), therefore could be used as an amendment in soils with acidic characteristics improving its productivity.
Acknowledgements: This work was supported by CONICYT-Chile, through FONDEF Project D07I1096. Furthermore, the authors thank Becas Chile for scholarship granted at one of them. Also acknowledgements at bilateral project 2009-145 CONICYT/CSIC y 2009CL0062 CSIC/CONICYT to found the stay made in the Institute IDAEA-CSIC, Barcelona, Spain.

Keywords: oat hull, pine bark, pyrolysis, biochar, specific surface area, pH

8th International Conference on Fate and Effects of Pulp and Paper Mill Effluents

ORAL PRESENTATIONS

ECOTOXICOLOGY OF EFFLUENTS AND THE RECEIVING ENVIRONMENT

Ecotoxicological impacts of pulp and paper mill effluents on carrying and assimilative capacities of plankton systems in Chilean rivers using a quantitative community to ecosystem level approach

B. Karrasch[1*], H. Cid[2], S. Woelfl[3], R. Urrutia[2], C. Valdovinos[2], R. Barra[2] and O. Parra[2]

[1]UFZ – Helmholtz Centre for Environmental Research, Brueckstr. 3[a], 39114 Magdeburg-Germany. Email: bernhard.karrasch@ufz.de; [2]Environmental Science Center EULA – Chile, University of Concepción. P.O. Box 160- C. Concepción-Chile; 3Universidad Austral de Chile, Instituto de Zoologia, Casilla 567, Valdivia-Chile.

For managing the environmental protection and welfare the quantification of stressor impacts and the elucidation of their relationships and interactions with ecological processes are important. Especially the development of management/remediation concepts and decision and policy making require parameter, which are comprising essential conversions of substances within aquatic systems and which are preferentially based on community to ecosystem level. Therefore we conducted first ecotoxicological investigations covering the quantification of carrying capacities of ecosystem services (1st step of the microbial self-purification, anabolic and catabolic processes of organic matter conversions, bioproductivity) of river plankton, subjected to sewage disposals from residentiary pulp and paper mills. For this kind of risk assessments we applied approved ecomicrobiological and planktological methods.

For nearly all microbial standing stocks (abundance and biomass) and activities we determined enhanced concentrations/activity rates in the effluent and in the river below the pulp mill discharge. This indicates that bacteria, which were adaptive to the extreme conditions within the process water of this industry, could use effective the herein abundantly available inorganic and organic nutrients. Mainly inhibiting impacts of the effluents were quantified for standing stocks of heterotrophic flagellates and primary production. Already after 1 to 5 kms below the discharge point the impacts of the effluents were significantly reduced or no longer detectable.

Direct ecotoxicological assessments of pulp mill effluents and their impacts on plankton systems exhibited clear inhibiting and stimulating effects on ecosystem services. The results are discussed within the context of the ecotoxicology of these effluents on microbial ecology.

Keywords: assimilative capacities, bacterial production, carrying capacities, ecotoxicology on community level, extracellular enzyme activity, primary production.

Main pulp and paper mill toxic internal effluents, its impact on WWTP and treated effluent quality

T. Heid-Furley and J. Lombardi

APLYSIA Tecnologia para o Meio Ambiente, Vitória, Brazil. E-mail:tatiana@aplysia.com.br

Pulp and paper mill effluents are composed of a wide chemical compounds variety from different sectors of the plant, which together can be toxic to the wastewater treatment plant (WWTP) microorganisms and to the receiving body. To reduce the ecotoxicity and minimize the impacts of pulp and paper effluents on the aquatic ecosystem is important to: identify the inmill toxic sources, estimate the toxic load contribution from each current process to the total, making possible in the future to manage them and reduce the impacts on WWTP, reduce WWTP costs and improve the overall quality of the effluent treated. This paper presents the APLYSIA experience characterizing fifty-nine inmill effluents from six pulp and paper Brazilian mills concerning their ecotoxicity, their effects on WWTP microorganisms and final treated effluent quality.

The results showed high toxicity contribution from contaminated condensates (13.559.322 UTm3/day), bleaching (276.498 UTm3/day) and paper machine (405.157 UTm3/day). These effluents have caused general effluent toxicity increasing, impacts on the bacterial flocs formation and protozoan survival from WWTP, reduction in BOD and COD removal efficiency, and increasing in suspended solids and turbidity concentration in the final treated effluent.

Keywords: Effluents, pulp and paper, ecotoxicity, wastewater treatment plant.

A comparison of effluents and wood feedstocks from pulp and paper mills in Brazil, Canada and New Zealand and their ability to affect fish reproduction: Part I - chemical profiling and in vitro assessments

C. Milestone[*1,2], R. Orrego[3], P. Scott[1], A. Waye[4], M. Kohli[2], B. O'Connor[5], B. Smith[6], H. Engelhardt[6], D. MacLatchy[1], V. Trudeau[4], T. Kovacs[5], T. Heid-Furley[7], A. Slade[8], M. Servos[6], D. Holdway[3] and M. Hewitt[2]

[1]Biology Department, Wilfrid Laurier University, Waterloo, ON, Canada, [2]National Water Research Institute, Environment Canada, Burlington, ON, Canada, [3]Faculty of Science, University of Ontario Institute of Technology, Oshawa, ON, Canada, [4]Department of Biology, University of Ottawa, Ottawa, ON, Canada, [5]FPInnovations-Paprican, Pointe-Clare, QC, Canada, [6]Department of Biology, University of Waterloo, Waterloo, ON, Canada, [7]Aplysia Institute, Vitória, ES, Brazil, [8]Scion, Rotorua, New Zealand.

Pulp and paper mill effluents have the potential to affect fish reproduction; however, clear relationships between industrial manufacturing processes, effluent treatment types and specific effects remain unclear. Comparisons between studies are difficult due to varied approaches involving different fish species, experimental conditions, and outcome measures. Furthermore, mill effluent compositions are chemically complex and vary with feedstock, process type, and treatment type. Our aim is to compare the effect(s) of feedstocks and final effluent from various mills on fish reproduction. Final effluents and wood chips from mills of representative process types and tree species were sampled and extracted. Here, we present results from both chemical profiling and in vitro testing of the extracts.

A filtration/solid phase extraction (SPE) regime was developed to extract wastewater samples producing four fractions. Woodchips were sequentially extracted to produce three fractions. Samples were analysed using GC-MS for known compounds, with GC Index (GCI) values indicative of extracted material for each fraction also determined.

Androgenicity and estrogenicity were tested using appropriate assays. Androgenicity was strongly associated with effluent non-polar SPE fractions, and showed a strong significant correlation to GCI. Little distinction was revealed between mills. Softwood extracts revealed high androgencity and strong correlation to GCI, while hardwood extracts showed minimal activity. All extracts showed minimal estrogenicity.

Extracts were also evaluated for their effect on two neurotransmitter enzymes involved in reproduction; monoamine oxidase (MAO) and glutamic acid decarboxylase (GAD). The greatest effluent extract activity was associated with the most polar fractions for both enzymes, with little distinction between mills. Weak significant correlations were observed between both polar and non-polar extracts. Woodchip fractions showed activity throughout all fractions, with little distinction between wood type/source.

Keywords: GC Index, Endocrine, Androgenicity, Estrogenicity, Neuroendocrine, MAO, GAD.

Evaluation of kraft wood pulp mill secondary effluent toxicity in Engraulis ringens and Strangomera bentincki embryos utilizing a developmental toxicity test

A. Llanos-Rivera*[1,2], , L. Castro[1,3], P. Vasquez, J. Silva[1,4] and E. Bay-Schmith[1,4]

[1]Programa Investigación Marina de Excelencia (PIMEX). Universidad de Concepción. [2]Unidad de Biotecnología Marina. Facultad de Ciencias Naturales y Oceanográficas, Universidad de Concepción (alllanos@udec.cl) [3]Laboratorio de Oceanografía Pesquera y Ecología Larval. Departamento de Oceanografía. Universidad de Concepción. [4]Laboratorio de Bioensayos, Departamento de Zoología, Universidad de Concepción.

The anchoveta, (Engraulis ringens) and sardine (Strangomera bentincki), are coastal pelagic species found from central (29°S) to southern Chile (42°S). During their peak spawning season, winter to early spring, high egg concentrations are found in an important spawning area located off the coast of central southern Chile. Close to this spawning area, a new offshore kraft wood pulp mill effluent discharge system has been installed, raising considerable social concern regarding the fate of the anchoveta and sardine spawning area. Studies in Chile on the potential impact that industrial effluents might have on local pelagic species, are scarce. Accordingly, the aim of this study was to evaluate the direct effect of the kraft wood pulp secondary effluent on the early stages of the aforementioned species. In a first phase, using a reference toxicant, we validated and standardized the anchoveta embryos from the natural environment as a test organism for ecotoxicological evaluations considering egg mortality as the endpoint. In the second phase, utilizing the embryo tests, we assessed the effluent toxicity in a dilution series of salinity adjusted Kraft wood pulp secondary effluent using anchoveta and sardine eggs collected from the natural environment. During the late winter-early spring spawning seasons between 2007 and 2010 we carried out a series of short cruises to the coastal area off Talcahuano. The eggs collected from the field were separated from the plankton less than 2 h after collection and the early developmental stages (phase III) were utilized in bioassays. Depending on the availability of eggs in the plankton, five eggs of each species were transferred to wells with freshly prepared test solutions and placed in temperature-controlled baths (12°C) under a 12 h light: 12 h dark photoperiod. The effluent stock solution salinity was set at 34 psu using hypersaline brine or synthetic sea water salt. Additionally, two extreme Potassium Dichromate concentrations (300 mgL-1 and 38.9 mgL-1) were included to check constancy in embryonic sensitivity. At hatching time (96 hrs), all rearing containers were checked and the number of dead embryos registered. The data were expressed as egg mortality and ANOVA was used to compare egg mortality between the effluent dilutions and controls. Our results from a series of seven bioassays performed with anchoveta embryos showed that three did not exhibit significant differences in the number of dead eggs between the effluent dilutions and controls ($P>0.05$ mortality mean 13.7% ± 7.3). In the rest of the bioassays significant differences were only observed between the mortality in the control group (mean 22.1% ± 9.3) and the treatment with 100% effluent (mean 61.1% ± 27.8). These results indicate that embryos sensitivity to the effluent was variable among batches corresponding to different sampling dates, probably coincident with egg quality seasonal variability reported for anchoveta in this spawning area. With sardine embryos, four bioassays were mounted and the results indicated no differences in mortality between the control group (mean 5.6 % ± 4.2) and secondary effluent dilutions (mean 5.8% ± 10.4). Based on all the results obtained we concluded that: i) effluent dilutions present in the environment (>1%) did not induce an increase in mortality of sardine and anchoveta embryos, ii) only the 100% effluent treatment produced an increase in anchoveta embryo mortality and this was coincident with the period in which embryo quality decreases. This information highlights the importance of assessing the effluent effects in each species separately and also of the importance of taking into account the complete reproductive cycle.

Keywords: Secondary effluent, embryos, toxicity test, Engraulis, Strangomera

Aquatic biology research for the forest products industry: an overview of recent studies and future directions

C. A. Flinders, W. Streblow, D. L. Borton, J. Beebe, D. McLaughlin, E. Schilling, P. Wiegand

National Council for Air and Stream Improvement, Inc. (NCASI)

The National Council for Air and Stream Improvement's (NCASI) aquatic biology program was initiated in 1973 to address questions related to the effects of pulp and paper mill effluents on fish; initially through the use of experimental stream mesocosm studies in the southeastern northwestern United States. Since that time, the program has expanded to include laboratory studies using daphnids, marine invertebrates, algae, and fish in both short-term to full life cycle evaluations. Research tasks focusing on the effects of effluent exposure, identifying changes in biological responses associated with advances in effluent treatment, and evaluating, performing and auditing bioassay methods sometimes mandated for regular use in monitoring mill effluent discharges. In recent years, the focus of research efforts has grown to incorporate large-scale, in-stream studies to examine the response of abiotic and biotic endpoints to effluent exposure in natural systems. The Long-term Receiving Water Study (LTRWS) explores aquatic communities across large spatial and temporal scales in four US effluent receiving streams to evaluate the role of effluent exposure and watershed characteristics in patterns of fish, macroinvertebrates, periphyton, water quality, and habitat variables. Efforts to develop models and tools to extrapolate the results of this and other in-stream studies to other systems are underway, as are studies integrating aquatic biology endpoints into forestry research examining stewardship and best management practices. This presentation will describe recent research efforts in greater detail and summarize key findings, as well as provide an outline of expected NCASI's aquatic biology research program directions in the short and medium terms.

Keywords: Long term monitoring, aquatic biology endpoints.

REPRODUCTIVE EFFECTS / ENDOCRINE DISRUPTION

Pulp mill effluents and pulp mill feedstocks as neuroendocrine disruptors

A. Waye, J. Guerrero-Analco, A. Saleem, M. Taylor, J. Arnason and V. Trudeau

Department of Biology – Ottawa, University of Ottawa. 30 Marie-Curie, Ottawa, ON., K1N 6N5, Canada. E-mail: awaye033@uottawa.ca.

Pulp mill effluents (PME) inhibit spawning in laboratory fathead minnow and suppress luteinizing hormone (LH) in wild white sucker. We hypothesized that chemicals in PMEs are disrupting neuroendocrine pathways that control reproduction, and that these chemicals are produced by trees used as pulp feedstocks. In vitro studies by our lab have shown the potential for PME and feedstocks to disrupt the neuroendocrine pathways responsible for spawning and LH release. GABA is an important neurotransmitter that stimulates LH release. Glutamic acid decarboxylase (GAD) is the enzyme that synthesizes GABA. We used bioactivity-guided fractionation using an in vitro GAD assay of a biologically treated PME and extracts of balsam fir (Abies balsamea, a Canadian pulp feedstock) to characterize neuroactive fractions and identify putative GAD-inhibiting compounds. Ninety litres of secondary effluent was concentrated by reverse osmosis and extracted by liquid-liquid extraction. Ethylacetate (EtOAc) extracts were inhibitory to GAD (54.4 – 76.2%). When we screened Canadian conifers for in vitro GAD inhibition, EtOAc extracts of balsam fir were identified as active (30.5% GAD inhibition). (+)-lariciresinol was isolated and identified (using silica column chromatography, preparative-HPLC, and NMR analysis), in an active (39.6% inhibition) fraction of a balsam fir EtOAc extract. Active sub-fractions (44.9 - 49.9% inhibition) of this fraction contained unidentified compounds to be further fractionated and isolated for identification and the tested for GAD inhibition. Purified (+)-lariciresinol was not inhibitory to GAD. These findings demonstrate that many neuroactive compounds are present in a pulp and paper feedstock.

Keywords: neuroendocrine disruption, glutamic acid decarboxylase, secondary-treated pulp mill effluent, luteinizing hormone, reproduction, bioassay-guided fractionation

Embryotoxicity of Chilean pulp mill effluent extracts in Japanese medaka (Oryzias latipes)

R. Orrego, J. Guchardi and D. Holdway

Aquatic Toxicology Laboratory, Faculty of Science. University of Ontario Institute of Technology, Oshawa L1H 7K4, ON. Canada. rodrigo.orrego@uoit.ca

Effects of Chilean pulp mill effluent extracts (solid phase extraction of untreated, primary and secondary treated pulp mill effluents), along with steroid standards (testosterone and 17β-estradiol) and a wood extractive standard (beta-sitosterol) were evaluated on developing post-fertilized medaka fish embryos. Our research has included a waterborne semi-chronic exposure experiment using 24 hpf (unknown sex) d-rR (orange-red) mutant strain medaka embryos and two waterborne exposure experiments using 24 hpf (unknown sex) FLFII (female leucophore free) and 72 hpf FLFII (phenotypic sex-identified) mutant strain medaka embryos.

Chronic exposure of both 24 hpf d-rR and FLFII strain embryos showed similar delay in time to hatch and decreased hatchability with no significant egg or larvae mortality observed in all pulp mill extract exposed embryos. In contrast, significant early hatching and increased hatchability were observed in beta-sitosterol d-rR exposed embryos. High mortality was observed in all testosterone exposed embryos. Severe teratogenic responses were observed in medaka d-rR embryos in all treatments. Abnormalities included optical deformities (micro-opthalmia, 1 or 2 eyes) and lack of development of forebrains and hearts. Teratogenic responses such as axis malformation and pericardial edema were observed in all treatments using 24 hpf FLFII with a high percentage related to males identified by fluorescent leucophore under microscope.

Sex-related mortality and teratogenic effects were observed in 72 hpf FLFII medaka embryos. Axis abnormalities and pericardial edema were associated with all treatments and similarly to the 24 hpf embryo experiment, mostly related to males. High mortality associated with severe malformations was observed in male and female embryos exposed to testosterone. Furthermore, signs of later masculinisation were observed in females exposed to testosterone where a slight fluorescence of leucophores was detected before death. Differences in the severity of teratogenic effects seem to be related to the unique phenotype of each medaka mutant strain (d-rR and FLFII).

Overall, our research has indicated that Chilean pulp and paper mill extracts caused medaka embryotoxicity (post-fertilized embryos) irrespective of the effluent treatment and different mutant strains used. The effects were mainly associated with delayed time to hatch, decreased hatchability, and sex-related teratogenesis and mortality.

Keywords: Pulp and paper mill extracts, Chile, embryotoxicity, teratogenic effects, Japaneses medaka

Comparing responses of sentinel fish species before and after the shutdown of a Kraft pulp mill in northern Ontario, Canada

C. Arciszewski* and K. Munkittrick

Canadian Rivers Institute - University of New Brunswick, 100 Tucker Park Road, Saint John, New Brunswick, Canada.
Email: tim.arciszewski@gmail.com

Studies on the impacts of pulp mills have been conducted at many sites in Canada since the late 1980s. Many of these studies have shown depressed circulating steroids and reduced gonad size. Temporary shutdowns and process improvements were often associated with moderate recovery of fish sentinels. In Jackfish Bay, Ontario, short-term shutdowns often elicited recovery of endpoints with high turnover, like mixed function oxygenases (MFOs) in the liver. A shutdown lasting 18 months also showed some reversal in the effects on fish, but not all impacts disappeared. Since 2000, the Canadian pulp and paper industry has contracted by approximately 50 %. With the permanent closure of many of these facilities comes the opportunity to more thoroughly study the recovery of ecosystems previously exposed to pulp mill effluents. At one kraft mill in Smooth Rock Falls, Ontario, Canada (which was closed permanently in July 2006), there was extensive work done while the mill was operating including collections by Environment Canada (Moose River Project; 1991-1997) and four cycles of Environmental Effects Monitoring (EEM; 1995, 1998, 2002, 2005). The existing dataset includes performance data (body condition, growth rate, age, gonadosomatic index (GSI), liver somatic index (LSI), liver enzymes (mixed function oxygenases; MFO), circulating sex steroids, and fecundity) from white sucker (Catostomus commersoni) collected in 1993-1998, 2002, and 2005. During the operation of the mill, strong and recurrent effects were found in female gonad size; significant ($p<0.05$) effects on GSI of female fish ranged from -5 to -38% compared to an upstream reference site. Given the impacts of the previous discharge of now-closed mills on fish reproduction, we have initiated studies to examine the recovery of the Mattagami River which received the pulp mill effluent from the Smooth Rock Falls mill. To better understand recovery processes and their potential rate, this study will use the same sentinel fish species (white sucker), study locations, and methodology to compare the operational (1993-2005) and post-operational (2011) datasets.

Keywords: Shutdown, monitoring, fish populations, reproduction, recovery, white sucker.

EFFLUENT CHARACTERIZATION / TOXICITY IDENTIFICATION EVALUATION

Investigation on color formation from pre and post coagulation treatment of Pinus radiata Sulfite pulp mill wastewater using nutrient limited aerated baffled stabilization basin treatment

R. Lewis[1], J. Van Leeuwen[1], A. Averson[1], and C.W. Chow[1,2] and M. Jimenez[1]

[1]*Centre for Water Management and Reuse, School of Natural and Built Environments, University of South Australia, South Australia, 5095; (Email: Rosmala.dewi@postgrads.unisa.edu.au; john.vanleeuwen@unisa.edu.au); 2Australia Water Quality Centre, South Australia, 5000.*

Treatment of wastewater from a Pinus radiata sulfite pulp and paper mill located in South Australia, using coagulation (1860 ppm of Alum) followed by retention in nutrient limited aerated stabilization basins (ASBs), was investigated. The aim was to determine a treatment process that maximizes BOD/COD removal, minimizes color formation and elucidate the molecular weight (MW) compounds that could contribute to color formation after the ASB treatment.
Two baffled ASB lab scale simulations (3 tanks in series, operated for 84 days [3 HRT]), with two different types of influent wastewater were studied. The influent wastewater for simulation 'A' was non-coagulated wastewater, whilst simulation 'B' used coagulated wastewater mixed with an adjacent paper mill effluent. Despite the fact that coagulation used as a preliminary treatment was able to achieve a high removal of color456 and UV260 and 280 (80%), color456 increased (>100%) with a reduction in UV256 and 280 (~10%) for both simulations. However, both ASB simulations achieved high BOD (~96%) and COD (~ 60%) reductions.
The color increased in either simulation was due to the formation of different MW compounds ('A' 1000 to 100,000 Da and 'B' 600 to 5000 Da) and SUVA256 (> 2 cm/L/mg for 'A' and < 1.2 cm/L/mg for 'B'). When compared with field data, 'A' performed similarly in terms of color development. Adequate BOD/COD removal was demonstrated in both simulations even though the nutrient loading was low, however it was concluded that the associated microorganism growth led to color formation. This paper provides insight on the characterization of recalcitrant organic compounds causing color formation.

Keywords: Aerated stabilization basins, coagulation, Pinus radiata, colour formation, nutrient limited, MW.

Isolation and identification of ligands for the goldfish testicular androgen receptor in chemical recovery condensates from a Canadian bleached kraft pulp and paper mill

P. Scott[1,2], C. Milestone[1,3], S. Smith[1], D. MacLatchy[1] and M. Hewitt[3]

[1]*Wilfrid Laurier University, Waterloo, Ontario, Canada;* [2]*Smart Water Research Centre, Griffith University, Queensland, Australia;* [3]*Environment Canada, Burlington, Ontario, Canada*

This study builds on a series of investigations characterizing substances in chemical recovery condensates at a Canadian bleached kraft mill that depress sex steroid levels in fish. In the present work, incubations of condensate extracts with goldfish testicular androgen receptors (AR) were used to streamline the development of chemical fractionation methods. After applying a solid phase extraction (SPE) protocol developed to recover known condensate extractives, non-polar extracts of filtered solids prior to SPE were found to contain high levels of androgenic activity. Using normal phase HPLC, activity was recovered in one fraction, with AR ligands detected in 3 additional fractions. Gas chromatography mass spectrometry (GC-MS) analysis of the most active fraction confirmed the two most abundant components as the diterpenes manool and geranyl linalool. Manool exhibited a relative affinity for the AR that was 0.003 of testosterone and accounted for 25 and 14 % of the activity of the filtered solids and the most active HPLC fraction, respectively. Geranyl linalool exhibited no affinity for the AR. 3 additional diterpenoid families were tentatively identified as the principal components of the 3 other androgenic HPLC fractions. Final effluent was found to have a low level of androgenic activity, with <1 % attributable to manool. Androgens previously identified in mill effluents (progesterone, androstenedione, androstadienedione) were not detected. This study is the first to confirm non-steroidal cyclic diterpenes as androgenic at kraft mills and present a plausible mechanism for mill effluents to affect fish reproduction.

Keywords: Condensates, endocrine disruption, investigation of cause, androgenic acitivty

Chromophoric compounds from an Eucalyptus bleached Kraft pulp mill effluent: new insights from comprehensive two-dimensional liquid chromatography

R. Duarte and A. Duarte

WCESAM & Department of Chemistry, University of Aveiro, 3810-193 Aveiro, PORTUGAL. E-mail address: regina.duarte@ua.pt

Macromolecular organic compounds, essentially lignin derived, are the main waste products in pulp mill effluents [1,2]. They are responsible for the notorious colour of those effluents, which is one of the environmental problems associated to this type of industry. The knowledge of the molecular structure of those chromophoric compounds in pulp mill effluents has become, therefore, important to predict changes of the bulk properties of the dissolved organic matter in receiving waters and it can also be useful for choice of effluent treatment procedures. However, the task of characterizing these complex organic mixtures is a true analytical challenge, since no single technique provides adequate information regarding all their different distributions (e.g. molar mass, chemical composition, functionality).

In this work, a comprehensive two-dimensional liquid chromatography (2D-LC) method was established for the separation and further characterization of molecular structures in chromophoric dissolved organic matter from a bleached Kraft pulp mill effluent. Such methodology comprised the use of a reversed-phase (RP) column in the first dimension and a size-exclusion (SEC) column in the second dimension, coupled to three on-line detectors (ultraviolet, molecular fluorescence, and evaporative light scattering). By applying two different separation mechanisms in a single run it was possible to draw insights into how the molecular weight profile of the chromophoric compounds relate to their hydrophobic features. The 2D-LC simplified fractions were further characterized in order to understand how the structural information varies with polarity and molecular weight profile.

Keywords: Eucalyptus bleached Kraft pulp mill effluent; chromophoric compounds; two-dimensional liquid chromatography; molecular weight distribution; hydrophobic features; structural characteristics

Characterization and removal of carbon, color, phosphorus, nitrogen and odor from magnesium bisulfite pulp mill wastewater for the remediation of a large coastal shallow freshwater lake

S. Nothrop[1], J. van Leeuwen[1], R. Dewi[1], A. Everson[1], C. Chow[3], E. Plant[2] and R. Smernik[2]

[1]*Centre for Water Management and Reuse – Australia, University of South Australia P.O. Box 2471 – Adelaide, South Australia, 5001. E-mail: notsc001@mymail.unisa.edu.au ;* [2]*Soils Group, School of Agriculture, Food and Wine, University of Adelaide;* [3]*Australian Water Quality Centre, SA Water, 250 Victoria Square, Adelaide, South Australia 5000*

The pulp and paper mill wastewater treatment processes developed in this research provide cost effective options for point source reductions of pollutants, for the control of environmental conditions within a large coastal shallow freshwater lake, Lake Bonney south east (SE), where phosphorus is the limiting nutrient for eutrophication. Removals of carbon (DOC), color, total phosphorus (TP), total nitrogen (TN), and odor from the wastewaters of a magnesium bisulfite (Magnefite) pulp mill by alum coagulation were studied utilising jar tests and pilot plant operation. Removals of 88% TP, 77% true colour, 39% TN and 28% DOC were achieved at an alum dose of 500 mg/L at pH of 4.3. Alum in combination with a polyamine was also assessed by jar tests and pilot plant operation, for removals of carbon (DOC), color, TP, TN, and odor from the wastewaters of a Magnefite pulp mill. Epichlorohydrin dimethylamide (EPI-DMA) was the polyamine used. The combination of 500mg/L of aluminium sulphate and 5 mg/L of EPI-DMA increased the coagulation pH to 5.2. Removals of 83.1% TP, 79% true colour, 51% TN, and 28% DOC were achieved. The increased TP removal was attributed to an increased coagulation pH which reduced the amount of soluble phosphorus in the final treated wastewater. The alum and polyamine combination resulted in the removal of significant odor from the Magnefite wastewater. When alum was used alone the odor remained. Subsequent sludge dewatering trials revealed the odour was absent in the alum-polyamine sludge and present in the alum sludge. The alum-polyamine sludge was noticeably harder, suggesting the presence of mercaptan hardeners. Dissolved aluminium in the alum sludge filtrate was 29 mg/L compared with 7 mg/L in the alum-polyamine sludge filtrate. Carbon-13 nuclear magnetic resonance (13C NMR) spectroscopy, high performance size exclusion chromatography (HPSEC) and ultrafiltration (with 30, 10, 5, 3 and 1 KDa membranes) were used to characterise the molecular weight fractions of the wastewater, treated water, sludge and water from Lake Bonney SE. In conclusion this research has provided feasible strategies for the management of industrial wastewater, which has provided up to 20% of the annual inflow to Lake Bonney SE since 1942, to prevent eutrophication and to enhance biodiversity within Lake Bonney SE.

Keywords: Coagulation, eutrophication, Epichlorohydrin Dimethylamide, odor removal, Lake Bonney SE

NUTRIENT ENRICHMENT

Prediction of in-river responses to various phosphorus-reduction strategies at a Canadian Kraft pulp mill through development of a river-simulation model

M. Davies[1], M. D. Lebo[2], P. Pagoria[3], E. Mannisto[4]

[1]Hatfield Consultants, North Vancouver, BC, Canada. E-mail: mdavies@hatfieldgroup.com; [2]Weyerhaeuser NR, Raleigh, NC, USA; [3]Weyerhaeuser NR, Highland Park, IL, USA; 4EKONO Inc., Bellevue, WA, USA.

When environmental effects have been confirmed in receiving environments of Canadian pulp and paper facilities, the federal Environmental Effects Monitoring (EEM) program requires an Investigation of Solutions (IOS) study, to identify ways to address the environmental effects identified in earlier phases of the EEM program. Enhancement effects on river biota have been observed in the Wapiti River of northern Alberta, and related to nutrients discharges to this river from both the Weyerhaeuser Cellulose Fibers FR Kraft pulp mill and the City of Grande Prairie sewage-treatment plant.

In Cycle Five of the EEM program, the Weyerhaeuser pulp mill at undertook an engineering investigation to identify and assess various process-modification scenarios that could reduce phosphorus discharged with final effluent, and developed a linked river-water-quality model, which used Monte Carlo simulation to explore potential effects of nutrient discharges from the pulpmill and municipal wastewater on the river's nutrient status over a range of river conditions. The design and outputs of this IOS modeling approach will be presented.

Keywords: Nutrient enrichment, pulp mill effluent, sewage treatment, phosphorus, nitrogen, periphyton

INVERTEBRATE / FISH RESPONSES

Lab fish tests to assess the effects of pulp mill effluents

J. Parrott[1], M. Hewitt[1], T. Kovacs[2], D. MacLatchy[3], P. Martel[2], B. O'Connor[2], M. van den Heuvel[4], G. Van Der Kraak[5] and M. McMaster[1]

[1]*Environment Canada, Burlington, ON joanne.parrott@ec.gc.ca.* [2]*FP Innovations, Pointe Claire, QC.* [3]*Wilfrid Laurier Univ, Waterloo, ON.* [4]*Univ of Prince Edward Island, Charlottetown PEI mheuvel@upei.ca;* [5]*Univ of Guelph, Guelph, ON gvanderk@uoguelph.ca.*

The effects of individual pulp mill effluents (PMEs) have been assessed in controlled laboratory tests with small bodied fish These tests have the benefit that they remove potential confounding factors that exist when testing fish collected from the field, such as issues of exposure to historically-contaminated sediments and other contaminant inputs (such as agricultural run off, municipal sewage treatment effluent input, etc.). Controlled laboratory fish exposures have been used to assess the potential for pulp mill effluents to affect gonad (ovary or tests) size, the number of eggs produced, or indicators of gonadal growth and function (such as measurement of reproductive steroids or vitellogenin). This presentation will review the findings of these short-term (less than 3 weeks) and long-term (4 month or more) reproductive indicator tests of fish applied to pulp mill effluents, and their potential use in addressing the issue of reproductive toxicity of PMEs. Short term exposures of juvenile rainbow trout, goldfish, fathead minnow, mummichog, and largemouth bass to PME have assessed vitellogenin (Vtg) production and sex steroid concentrations. Most tests showed decreased steroid levels in fish exposed to PMEs, mimicking the steroid reduction seen in wild fish. These tests also have shown that exposures of fish to PME results in changes in egg production, secondary sex characteristics, gonadosomatic index (GSI, a ratio of gonad weight to body weight), liver-somatic index (LSI, a ratio of liver weight to body weight), and growth. In North America, fish lifecycle tests (120 days or more) have usually been conducted with fathead minnows. Because the lifecycle test is meaningful, but costly, researchers are attempting to shorten the lifecycle test while still retaining the biologically-relevant endpoints. The common finding of reproduction as the most sensitive response to PMEin both short term and lifecycle tests holds promise for shortening the lifecycle tests, while still maintaining this important and meaningful endpoint. The fathead minnow adult reproductive assay has recently been tested with PMEs. As well, a similar assay with zebrafish is also being used to assess breeding. These and other fish tests have been applied to several PMEs, and used to determine where in the PME the causative compounds are arising. In conjunction with toxicity Identification and Evaluation (TIE) these efforts will identify compounds, sources, and eventually, ways to treat effluents to reduce impacts on fish reproduction.

Keywords: Fathead minnow, zebrafish, bioassays, reproduction, lifecycle, BKME, reproduction, liver size

Dynamics of fish community structure from Itata River in a section with the influence of Pulp mill effluent

N. Ortiz, A. Elgueta and E. Habit

Environmental Science Center EULA – Chile, University of Concepción. P.O. Box 160- C. Concepción-Chile. E-mail: neortiz@gmail.com

The assemblage of fish in freshwater system is considered as the highest trophic level in aquatic system. Therefore it is expected that the disruption to the environment due to the effect of industrial effluents on fish and other trophic levels such as phytobenthic algae and macroinvertebrates communities, have an effect on the structure of the fish fauna. Parameters were evaluated community diversity, richness, and similarity. Over four sample sites in the catchment area for periods of low flow (January and February) for 6 years (2006 to 2011). We discuss the scale dimension, and limitations of statistical methodology for the evaluation of the effect. Also emphasizes the need to study the responses of individuals through an experimental research.

Keywords: Industrial effluents, Cellulose, river itata, freshwater, fish, diversity, richness species

Reproductive and metabolic responses of Percillia gillissi and Trichomycterus areolatus populations after 5 years of bleached kraft mill effluent discharge and discharge divertion to sea.

G. Chiang[1,2], K. Munkittrick[2], M. McMaster[3], R. Urrutia[1], C. Concha[1] and R. Barra[1]

[1]Center for Environmental Sciences EULA-Chile, University of Concepción, Barrio Universitario s/n, P.O. Box160-C, Concepción, Chile Ph.: +56-41-2203293, Fax: +56-41-2207076, email: gchiang@udec.cl; [2]Canadian Rivers Institute, University of New Brunswick, St. John, NB, Canada; 3Environment Canada, Burlington, ON, Canada

Between 2007 and 2011, we monitored the reproductive status of wild populations of native fish in a basin of central Chile, the potential effect of exposure to effluent discharges from ECF pulp plant cellulose with a modern production and processing system (greenfield mill). We have confirmed endocrine disruption at different levels of biological organization in populations of two species of native freshwater fishes of Chile (Percilia gillissi and Trichomycterus areolatus) downstream the discharge during periods of 2007 to 2009. Depending on the seasonality of monitoring have shown an induction or inhibition of the production of sex steroids, gonadal maturation and histological level increased or decreased gonadosomatic index (GSI) during periods of recrudescence and spawning. In January 2010, the plant began to discharge into the sea, but due to the earthquake of February 27, 2010, the plant discharged into the river again briefly. Despite there was a high variability in the size of the gonad of P. gillissi, we observed a decrease in gonad size, along with induction of EROD activity downstream of the discharge. T. areolatus showed no differences with the reference sites for the same period. During periods of 2010 and 2011, there have been recoveries of populations downstream of the discharge, especially in T. areolatus. While there is a difference in the sensitivity and resilience of the species, is showing greater resilience in T. areolatus. Both species appear to exhibit an overstimulation of the reproductive system during exposure to effluents from pulp, which carries a positive and/or negative feedback of the reproductive system, dependent on the gonadal maturation period to be monitored. These changes have not reached a critical threshold of disturbance, as there is evidence of resilience in both species, by suspending the discharge of effluents from pulp to its habitat.

Keywords: Freshwater fish reproduction, Chile, pulp mill effluents, long term monitoring

INVESTIGATION OF CAUSES (IOC)

Androgens and fifth effect evaporator condensates do not cause similar responses in mummichog (Fundulus heteroclitus)

R. Rutherford[1], P. Scott[1,2], C. Milestone[1,2], A. Lister[1], M. Hewitt[2] and D. MacLatchy[1]

[1]*Laurier Institute of Water Science & Canadian Rivers Institute, Wilfrid Laurier University, Waterloo, Canada;* [2]*Aquatic Ecosystem Protection Research Division, Environment Canada, Burlington, Canada*

Studies at Irving Pulp and Paper Ltd, a bleached kraft mill in Saint John, New Brunswick, Canada, have successfully identified an in-mill waste stream (5th effect evaporator condensates) as containing endocrine disrupting substances (EDSs). These EDSs affect levels of reproductive steroids in a native estuarine fish, the mummichog or killifish (Fundulus heteroclitus). To chemically isolate suspect EDSs, a solid phase extraction protocol was used in 2009 to generate four chemically-distinct fractions, along with residual condensates. Contrary to previous work, there was no significant response in fish exposed to 4% (v/v) whole condensates, making interpretation of specific fractions impossible. The most compelling explanation in the variability of condensate effects over time is change in the chemical make up of the condensates. Chemical fingerprinting of the condensate batch found a higher androgenic potency in the fractionation condensates than a previous exposure in 2009 that elicited plasma testosterone depressions in males at 4% (v/v). Mummichog were then exposed to model androgens methyltestosterone (100 and 1000 ng/L; MT) and dihydrotestosterone (10 and 100 μg/L; DHT). Plasma testosterone was significantly depressed by 10 μg/L DHT in males. Females had significant depression of plasma estradiol when exposed to both DHT concentrations and 1000 ng/L MT. Comparison of effects in fish exposed to pulp mill condensates and model androgens indicates androgens in the condensates may not be the source of reproductive endocrine effects in condensate-exposed mummichog. Model hormone studies of this type could help determine the classes of active compounds in reproductive-active effluents.

Keywords: Fifth effect condensates, androgens, Fundulus heteroclitus, endocrine disruption.

Investigation of impacts on white sucker (Catostomus commersoni) populations from Jackfish Bay to over 20 years of changes in pulp mill effluent discharges

K. Munkittrick[1], M. McMaster[2], L. Bowron[3], G. Tetreault[2] and M. Hewitt[2]

[1]Canadian Rivers Institute and Department of Biology, University of New Brunswick, 100 Tucker Park Road, Saint John, NB, Canada . Tel.: +1-506-648-5985; fax: +1-506-648-5811; [2]Environment Canada, 867 Lakeshore Road, Burlington, ON, Canada; 3Dept. Biology, University of Waterloo, Waterloo, ON, Canada

The impacts of pulp mills effluents on fish and fish habitat have been studied at Jackfish Bay, Lake Superior, Ontario, Canada since the late 1980's. Jackfish Bay has received the effluent from a large bleached kraft pulp mill (BKM) in Terrace Bay, ON since 1948. The site is isolated from other sources of contamination, industrial or municipal effluents. Ongoing studies conducted since 1988, showed that white suckers (Catostomus commersoni) had a number of reproductive alterations, including delayed maturity, decreased gonadal size, decreased expression of secondary sexual characteristics, and depressed levels of circulating steroids. Reductions in gonad size and steroid levels persisted until the pulp mill closed its hardwood line in the late fall of 2005, and ceased all effluent discharge from its softwood line from February 2006 to September 2006. Fish were collected in the spring and fall of 2006 during shutdown, and then again 8 months after the mill reopened. The mill subsequently closed for >16 months, and fish were sampled after the mill reopened; sampling continued in 2011. This paper will discuss the pace of recovery after closures, and continuing concerns about the impacts of pulp mill effluents on fish reproduction.

Keywords: Recovery, Jackfish Bay, Fish reproduction, Mill shutdown.

Progress status of a collaborative research project on the causes and solutions for effects on fish reproduction in Canada

P. Martel[1], T. Kovacs[1], B. O'Connor[1], L. M. Hewitt[2], D. MacLatchy[3], M. McMaster[2], J. Parrott[2], M. van den Heuvel[4] and G. Van Der Kraak[5]

[1]FP Innovations, Pointe-Claire, QC, pierre.martel@fpinnovations.ca, tibor.kovacs@fpinnovations.ca, brian.oconnor@fpinnovations.ca; [2]Environment Canada, Burlington, ON, mark.hewitt@ec.gc.ca, joanne.parrott@ec.gc.ca, mark.mcmaster@ec.gc.ca; [3]Wilfrid Laurier University, Waterloo, ON, dmaclatchy@wlu.ca; [4]University of Prince Edward Island, Charlottetown, PEI, mheuvel@upei.ca; [5]University of Guelph, Guelph, ON, gvanderk@uoguelph.ca

The first five cycles of the Environmental Effects Monitoring (EEM) program in Canada identified two main environmental issues related to mill effluents. For benthic invertebrates, the evidence pointed to enrichment. In the case of fish, evidence of enrichment was accompanied by metabolic disruption as the larger fish had smaller gonads and this led to concerns about overall reproductive capacity. When effects are identified, the EEM program calls for Investigation of Cause (IOC) and Investigation of Solution (IOS) studies. To study effects in fish a research consortium was formed in 2005. To date, significant progress has been made. A laboratory test assessing the effect of effluents on egg production by the fathead minnow (Pimephales promelas) was found to be a reliable tool for monitoring effluent quality. Application of this test in surveys of mill effluents and monitoring at a kraft mill provided new leads for identifying mill operating conditions that could influence fish reproduction. During cycle 6 EEM our approach is to extend the monitoring to mills representative of major pulp and paper production processes across Canada. Integrated into the overall design of the study is the capacity to take advantage of opportunities for additional testing if results suggest a promising avenue of investigation or if a mill is implementing a change in operating conditions that could influence effluent quality. From this accumulated data we will identify a series of recommendations for mill operators that clearly result in improved effluent quality and a reduction in observed effects on fish reproduction.

Keywords: Environmental Effects Monitoring, pulp mill effluent, Pimephales promelas, gonads, reproduction

MONITORING PROGRAMS

The use of easy measurable variables in the sediment to assess impact in marine sediments affected by organic wastes (1978-2010)

X. Niell, R. Carmona and M. Ruiz

[1]*Department of Ecology. Faculty of Sciences. Campus de Teatinos. University of Málaga 29071. Málaga, Spain*

The assessment of impact in sediments is feasible fitting easy measurable variables like organic matter content, chlorophyllic pool pigments and elementary analysis of C and N. From these data several indexes reveal clearly the quality of the sediment in terms of potential primary production and flow of energy and organic matter degradability in sedimentary microecosystems. The sampling campaigns carried out since 1978 in Placeres (Pontevedra, Spain) allow to compare data after several improvements in the environmental management, and to compare the current state of the system with values obtained in other sites non affected by pulp-mill factories wastes. Data have been used to perform an useful model to assess impact of the organic wastes on sediments, to be applied easily and expressed as percentage of disruption related to the control reference system.

Keywords: Sedimentary ecosystems, Chlorophyll a, C/N, organic matter, dilution-diffusion functions, impact assessment

Monitoring of fish community in the recipient of a greenfield pulp mill at River Uruguay during 2007-2010

J. Tana[1], F. Teixeira de Mello[2], I. Gonzalez[2], N. Vidal[2], A. D'Anatro[3], D. Larrea[4]

[1]ÅF-Consult Ltd, Rajatorpantie 8, FI-01600 Vantaa, Finland Ph + 358 4 3485644 email jukka.tana@afconsult.com [2]Grupo de Ecología y Rehabilitación de Sistemas Acuáticos, Departamento de Ecología y Evolución, CURE, Universidad de la República. Maldonado, Uruguay. [3]Departamanto de Ecología y Evolución. Facultad de Ciencias, Montevideo, Uruguay [4] Biologist Consultant

A new Greenfield kraft mill started its operation in Uruguay in November 2007. The mill uses the latest pulp processing technology and is also the first modern kraft mill in Uruguay discharging the treated effluent to River Uruguay. As part of the environmental impact assessment process the mill started baseline studies on water quality, river biology and fish studies in the future recipient and reference areas two years prior to the start up. The studies have continued as monitoring studies. The abundance, biomass and species richness of the fish community as well as the condition and exposure state of fish have been studied. The task has been a serious challenge because there are more than 150 fish species in Uruguay River basin and very little is known about their life history. This is the first time in River Uruguay when regular fish monitoring studies have been performed at the same sites and using the same method. These monitoring studies have also given the opportunity to study the effects of the pulp mill effluents on fish community and to compare the results to baseline studies prior to mill start up.

The studies have been performed two times per year in April and December using Nordic standard gill net series in two recipient areas and one reference area. Baseline studies indicated that the best species as an indicator species would be Bagre trompudo (Iheringichthys labrosus) because of its feeding character, number and size classes in test fishing catches and expected locality in distribution. The liver and gonad somatic indexes, condition factor and exposure state of this species have been included in the monitoring studies. This paper presents the fish studies of a three year monitoring period and the comparison to the situation prior to mill operation. The availability of the chosen indicator species will be also discussed in the paper.

Keywords: Fish studies, monitoring, pulp mill effluent, water quality, greenfield mill, Rio Uruguay

The Environmental Effects Monitoring Program for the Pulp and Paper Industry in Canada: National Patterns resulting in IOC for Reduced Investment into Reproduction

M. McMaster[1], J. Parrott[1], M. Hewitt[1], T. Kovacs[2], D. MacLatchy[3], P. Martel[2], B. O'Connor[2], M. van den Heuvel[4], and G. Van Der Kraak[5]

[1]*Environment Canada, Burlington, ON joanne.parrott@ec.gc.ca, mark.hewitt@ec.gc.ca, mark.mcmaster@ec.gc.ca;* [2]*FP Innovations, Pointe Claire, QC tibor.kovacs@fpinnovations.ca, pierre.martel@fpinnovations.ca, michael.paice@fpinnovations.ca;* [3]*Wilfrid Laurier Univ, Waterloo, ON dmaclatchy@wlu.ca;* [4]*Univ of Prince Edward Island, Charlottetown PEI mheuvel@upei.ca; 5 Univ of Guelph, Guelph, ON gvanderk@uoguelph.ca*

Environmental Effects Monitoring (EEM) is a scientific tool that assesses the effects of effluent from industrial or other sources, on fish, fish habitat or the human use of the fisheries resources. Pulp and Paper Environmental Effects Monitoring was Canada's first national EEM program, has been followed by Metal Mining and Municipal Wastes are in development. The Canadian government provides advice on existing regulations, approval of study designs, review of reports, maintains a national EEM database and conducts national assessments of the data. The cycles of monitoring and reporting started with Cycle 1 in 1996, Cycle 2 in 2000, Cycle 3 in 2004, Cycle 4 in 2007 and Cycle 5 in 2010. National Assessments of the Pulp and Paper EEM Program have been completed and have identified distinct response patterns. The national pattern identified consistently in fish is one of metabolic disruption where fish grow faster, store more energy but put less into reproductive development. A national project was formed to investigate the cause of this reproductive dysfunction consisting of government, academia and industry scientists. This presentation with overview the national responses in wild fish as well as some of the more detailed reproductive studies.

Keywords: EEM Canada, Pulp and Paper mill Effluents, Fish, Cycles 1-5

Monitoring of Pulp Mill industries in Chile: Reflexions and some recommendations

R. Barra[1] and G. Chiang[1,2]

[1]EULA Chile Environmental Sciences Centre, University of Concepción, Barrio Universitario S/N Concepción, ricbarra@udec.cl, Chile; 2 1Canadian Rivers Institute and Department of Biology, University of New Brunswick, 100 Tucker Park Road, Saint John, NB, Canada

Given the importance of the pulp mill industry in Chile and the variety of receiving environments receiving their liquid effluents, it is important to establish some minimal requirements of environmental monitoring according to the needs for environmental protection and the scientific progress. So far any mill has their own requirements for monitoring in their receiving environment, and this is understandable, since each mill have been approved in a process considering the project itself by the regulatory authorities and the elements behind the approval for the new developments. One of the major constraints of the existing monitoring requirements is the heavy rely on physico-chemical parameters, with a poor consideration/ interpretation of the biological endpoints selected for the monitoring program and the relatively poor design of such programs looking at only higher levels of organization i.e. community composition and species richness and even no considering potential confounding factors. On the other side the scientific evidence clearly shows that old problems such as dioxins releases and chlorinated chemicals in the effluents does not represent a threat so far, new challenges need to be faced such as the metabolic and endocrine disruption, observed in many different situations around the world and even in Chile. We have been working in the last years on the use of small fish as sentinels of environmental changes caused by pulp mill discharges in Chile, and in some experimental approaches that may be used in the pulp mill environmental control when the effluent is discharged into a river. However, some improvements are also needed in the effluents regulation, since the chemicals causing the observed effects are unknown so far, it is unpractical only base the controls in chemical analysis, a more comprehensive approach should be used, such as toxicity testing that may help to prevent problems in the receiving environment. In this work some recommendations will be indicated to face these new challenges in the Chilean case.

Keywords: Monitoring, aquatic impacts, endocrine disruption, regulations

COP´s space-temporal distribution in sea water, marine sediments and marine organisms in the coastal platform off Itata river discharge area, Octava Region, Chile

M. Salamanca[1 2 3], C. Chandia[2 3] and A. Camaño[4]

[1]Laboratorio de Oceanografía Química (LOQ). Universidad de Concepción; [2]Departamento de Oceanografía. Universidad de Concepción; [3]Proyecto de Monitoreo Marino Nueva Aldea (PROMNA). msalaman@udec.cl, Casilla 160-C. Concepción, Chile; [4]Gerencia Medio Ambiente, Seguridad y Salud Ocupacional. Arauco El Golf 150 Piso 11, Santiago

The AOX, EOX, dioxins and furans content obtained during five years monitoring program of Nueva Aldea Cellulose Industrial Complex, carried out to verify the operation of a submarine discharge in the coastal platform off Itata river mouth is presented. Sampling results every three months since May 2006 through November 2010, shows a seawater AOX average concentration between 0.015and 0.220 and ND (not detected) to 0.180 mg L-1 for subtidal and intertidal environments, respectively. The average seawater dioxins and furans content varies between ND to 0.0099 ng TEQ L-1 and ND to 0.0030 ng TEQ L-1, for subtidal and intertidal environments, respectively, dominated mainly by OCDD, OCDF, HpCDD y HpCDF congeners. The average sediment AOX content varies between 55.5 to 247.5 and 47.5 to 250.5 mg kg-1 for subtidal and intertidal environments, respectively. The evaluated organisms presented an average AOX content between ND to 116.0; 0.00 to 186.8; 0.73 to 127.9 and ND to 424.4 mg kg-1 and an EOX average content between ND to 7.3; 0.00 to 10.2; ND to 19.6 and ND to 2.2 mg kg-1 for gastropods, crustaceans, fishes and bivalves mollusks, respectively. The average dioxins and furans concentration for gastropods, crustaceans, fishes varies between ND to 0.0001; ND to 0.0044 and ND to 0.0062 ng TEQ g-1, respectively, dominated mainly by OCDD, OCDF, HpCDD y HpCDF congeners. It is concluded that the analyzed COP´s does not present any seasonal or spatial pattern allowing identification of any localized sources, instead, theses result strongly suggest that the presented values correspond to local background levels.

Keywords: COP´s, coastal zone, monitoring, Itata River, Chile

Biodiversity of the coastal zone off the Itata River mouth after 5 years of effluents release from a pulp mill

E. Hernández-Miranda, Á. Araneda, E. Díaz, R. Veas, K. Ibáñez, L. Lozano and R. Quiñones

Programa de Investigación Marina de Excelencia, Facultad de Ciencias Naturales & Oceanográficas, Universidad de Concepción, Chile. E-mail: eduhernandez@udec.cl

Marine coastal ecosystems are highly variable and their physical, chemical and biological dynamics operate on various spatial and temporal scales. In particular, the coastal zone off central Chile possesses the greatest surface area of continental shelf of the Chilean coastline, and is modulated mainly by the following factors: (1) meteorological seasonality, upwelling and natural hypoxia events on an intra-annual scale, and (2) El Niño and La Niña events on an inter-annual scale. Thus, the understanding of these factors is fundamental for the development of small-scale studies with a view to determine the distribution and abundance of resident marine species in the area. Within this oceanographic-ecological context, in the area surrounding the Itata River mouth, the discharge of cellulose effluents from the Nueva Aldea Pulp Mill (CFI-Nueva Aldea) has emerged as a new potentially modulating factor. These effluents have been discharged into the sea over approximately the past 5 years, in three stages: first, the discharge of effluents with tertiary wastewater treatment into the Itata River, 45 km upstream of the mouth (August 2006 - August 2009); second, direct discharge of effluents with tertiary wastewater treatment into the sea (August 2009 - January 2010); and third, direct discharge of effluents with secondary wastewater treatment to the sea (January 2010 - April 2011).

Ecological research in this coastal area was started in June 2006 by the Program of Marine Research of Excellence (PIMEX) of the University of Concepcion, with the objective of describing the general patterns, processes and mechanisms which determine the distribution and abundance of the species that inhabit the Itata River mouth area (the recipient area of effluents from CFI-Nueva Aldea) and the surrounding areas between Cobquecura to the north and Bahía Coliumo to the south, a total of approximately 400 km2. After almost 5 years of research, the record and quantification of species in the study area has allowed us to: (i) identify and quantify the richness-composition of species in the ecosystem, made up of 170 species, with the dominant classes Polychaeta, Malacostraca, Gastropoda and Actinopterygii, (ii) identify the spatio-temporal structure of soft-bottom communities, which reveals a distribution that is highly related to water column dissolved oxygen, (iii) identify key species for long term demographic studies, specifically the polychaetes Prionospio peruana and Mediomastus branchiferus, the crustaceans Cancer coronatus and Emerita analoga, the gastropods Nassarius sp and the fish Paralichthys adspersus and Aphos porosus, (iv) compare the ecological conditions in the area before and after the operation of the CFI-Nueva Aldea, with results to date that reveal no temporal trends but rather evidence important temporal variability and, (v) contribute to the knowledge of the responses by marine populations faced with strong environmental perturbations, such as natural hypoxic events that have resulted in massive mortalities of marine organisms, mainly in Coliumo Bay, and the negative effects of the earthquake and tsunami of February 27th 2010, which strongly modified the habitat and decreased the short term species richness and density in this coastal area off central Chile.

Acknowledgements. This research is part of the Programa de Investigación Marina de Excelencia (PIMEX) of the Faculty of Natural and Oceanographic Sciences of the University of Concepción funded by Celulosa Arauco y Constitución S.A.

Keywords: Diversity, population, community, Humboldt Current System

Biological baseline and monitoring of pulp mill activities at Uruguay River

M. Dabezies, G. Ferrari, J. Clemente, L. Boccardi and C. Saizar

Departamento Medio Ambiente. Laboratorio Tecnológico del Uruguay, Montevideo, Uruguay. E-mail: mdabez@latu.org.uy

The biological baseline and monitoring studies in the lower Uruguay River undertaken by the environmental impact assessment process of the Kraft UPM (ex Botnia) pulp mill are presented in this study. The samples were collected seasonally between 2006 & 2010 at three river sites: Nuevo Berlín (upstream), Fray Bentos (mill zone) and Las Cañas (downstream). Three transections, perpendicular to the coast line, with three points each one, were defined for every locations: littoral, central and channel (river zones). The physico-chemical parameters of water (in situ measurements and nutrients), sediments (granulometry, organic matter, nutrients, metals, EOX, PCBs, PAHs, dioxins and furanes) and biological studies (phytoplanktic, zooplanktic and zoobenthic communities) were analyzed. As an average, phosphorus in the water column presented values above those defined by the national norms, while organic pollutants (AOX, chlorophenols, resinic acids, phytosterols, dioxins and furanes) remained below the recommended international guidance values. The physico-chemical parameters (water and sediments) and the three communities showed significant temporal differences while physico-chemical sediments parameters varied also in the spatial scale ($p<0{,}001$). Zoobenthic was the only community that presented spatial differences between zones.

Keywords: Environmental impact, water quality, contaminant sediments, phytoplankton, zooplankton, zoobenthos

RIVER / MARINE ENVIRONMENT IMPACTS

A reassessment of Fucacean communities next to a cancelled waste in the Northwest of Spain. Effects of the improvement of the environmental protection technology

R. Carmona, X. Niell, M. Ruiz and R. Sánchez de Pedro.

Department of Ecology. Faculty of Sciences.Campus de Teatinos. University of Málaga. 29071 Málaga, Spain

The distribution of fucaceans is characteristic and well known in the Atlantic coasts of Western Europe. In 1976, biomass losses and the disruption in the zonation of fucaceans around a waste point of a pulp-mill factory situated in the Ria of Pontevedra (NW Spain) were published. After several efforts to improve the technological facilities to avoid environmental damages, the degree of recovery of those fucacean communities of the intertidal ecosystem after 12 years of waste management can be determined.

Keywords: Fucaceae, intertidal zonation, biomass, diversity, restoration of benthic communities' structure

Degradation capability of biopolymers in the coastal environment off central Chile (36.5 S)

M. H. Gutiérrez[1], S. Pantoja[1,2], P. Ampuero[3], E. Tejos[2] and R. A. Quiñones[1,2]

[1]Department of Oceanography and Center for Oceanographic Research in the eastern South Pacific, University of Concepcion, Concepcion, Chile; [2]Program of Marine Research of Excellence (PIMEX-Nueva Aldea), School of Natural and Oceanographic Sciences, University of Concepcion, Concepcion, Chile; [3]Graduate Program in Oceanography, Department of Oceanography, University of Concepcion, Concepcion, Chile

The response of the coastal ocean influenced by both river discharges and inputs of photosynthetically derived organic carbon product of upwelling, was evaluated by estimating rates of microbial hydrolysis of macromolecules with the goal of estimating the degradation capability of the coastal ecosystem off central Chile. Extracellular enzymatic hydrolysis in seawater was dominated by aminopeptidase activity on substrate MCA-leu (1.2 to 182nMh−1), which surpassed that of substrates MUF-glu (0.08 61nMh−1) and MUF-cel (0.15–6.8nMh−1), corresponding to activity of glucosidase and glucanase from the cellulase enzymatic complex respectively, with the fastest rates measured during spring-summer. In riverine waters, extracellular enzymatic hydrolysis remained within the range of 45 to 131nMh−1 for MCA-leu and ca. 20nMh−1 for glucosidic substrates, year-round. Contrary to the observed for the marine wate column, surface sediment extracellular enzymatic hydrolysis of MCA-leu (0.04 to 6.13 nmol gdw−1 h−1) was in the same order of magnitude as the rates observed for MUF-cel (0.004 to 2.58 nmol gdw−1 h−1) and MUF-glu (0.007 to 10.5 nmol gdw−1 h−1). Moreover, hydrolysis was characterized by higher rates during winter compared with spring-summer in the coastal and estuarine zone. The five years of data allowed us to construct a carbon budget for winter and spring-summer in the coastal area adjacent to the Itata River discharge. Size-fractioned experiments determined that over 90% of extracellular enzymatic activity occurred in the size classes dominated by filamentous fungi during periods of high autotrophic biomass in surface waters of the upwelling ecosystem off central Chile.

This research was funded by the Programa de Investigación Marina de Excelencia, PIMEX-Nueva Aldea (University of Concepción and Celulosa Arauco y Constitución S.A.), and the Center for Oceanographic Research in the eastern South Pacific (COPAS).

Phytosterols distribution in coastal and river sediments in central-southern Chile

L. Saavedra[1,2], R. Quiñones[1,2,3] and J. Becerra[4]

[1]Programa de Investigación Marina de Excelencia, Facultad de Ciencias Naturales y Oceanográficas, Universidad de Concepción, Chile; [2]Programa de Doctorado en Oceanografía, Departamento de Oceanografía, Universidad de Concepción, Chile; [3]Centro de Investigación Oceanográfica en el Pacífico Sur Oriental (COPAS), Universidad de Concepción, Chile; [4]Laboratorio de Química de Productos Naturales, Facultad de Ciencias Naturales y Oceanográficas, Universidad de Concepción, Concepción, Chile.

Concentrations of sterols and lipid biomarkers (fatty acids, fatty alcohols, alkanes) in sediments from four coastal zones (the Gulf of Arauco, the Bio Bio River canyon, the shelf off the Itata River mouth, Coliumo Bay) and two rivers (Itata and Bio Bio) in central-southern Chile, were determined by capillary gas chromatography-mass spectrometry, with the aim of characterizing the distribution of phytosterols and their possible sources. Of the 17 sterols found in marine stations, the most abundant was cholesterol (~60%, 0.06 – 6.73 μg g-1). Of the 10 sterols from river sediments, the most important was β-sitosterol (0.01 - 2.33 μg g-1). The sources of sedimentary organic matter at each station were determined using different lipid biomarkers and sterol ratios. Some stations in coastal areas adjacent to the Itata and Bio Bio rivers had β sitosterol of clearly terrestrial origins. On the other hand some oceanic stations had β-sitosterol of phytoplankton origin as indicated by the presence of dinosterol. The Bio Bio River and its mouth had a wider diversity of sterols and lipids and high levels of cholesterol and epicholestanol, which are likely to be related to the presence of sewage from large cities. The level of sterols found off central southern-Chile seems to be much lower than levels that could be reached in other portions of the Humboldt Current System, such as Peru, under natural conditions. Several of the sampling stations with higher β-sitosterol content were associated with river mouths and pulp mill industries, but others were far from those influences.

This research is part of the Programa de Investigación Marina de Excelencia (PIMEX) of the Faculty of Natural and Oceanographic Sciences of the University of Concepción, funded by Celulosa Arauco and Constitución S.A.

Keywords: Sterols, phytosterols, terrestrial biomarkers, river sediments, coastal zone, fatty acids, Humboldt Current Systems

More than fourty years of coevolution in a multiple environmental system. Factories, cities and estuarine ecosystems. Is it possible a steady-state in the whole?

X. Niell[1], C. Casas[2], R. Carmona[1] and M. Ruiz[1].

[1]Department of Ecology. Faculty of Sciences. Campus de Teatinos. University of Málaga. 29071 Málaga, Spain. [2]ENCE Pontevedra. Marisma de Lourizan s/n. 36153 Pontevedra, Spain

The Ria of Pontevedra is one of the highest productive estuaries sited in front of an upwelling intense area in the Northwest coast of the Iberian Peninsula. Environmental changes in this area have been recorded along half a century in many ways and comparisons could distinguish the influence of the industrial activities, their wastes and the effect of the increase in population along the coast of this estuary.

The expected steady state of the system is evaluated and discussed in this study, as well as the changes happened in the macroecosystem after several measurements taken in order to manage the waste of the industries and the built-up urban areas in a rich natural eutrophic estuary forced in its production by the oceanic circulation in the area, which is responsible for the sustainment of a high primary production, as well as the exploitation of a rich shell fish area. Some suggestions to manage this complex system are pointed out in order to support the decision-makers in environmental policy in such a complex macrosystem.

Keywords: Multiphocal wastes influences. Coexistence of stuctures. Estuarine evolution. Sustainability

Declining levels of dioxins in sediment and fish in the receiving waters of two Swedish bleached kraft pulp mills

M. Karlsson and J. M. Malmaeus

[1]*IVL Swedish Environmental Research Institute, P.O. Box 210 60, SE-100 31, Stockholm, Sweden. E-mail: magnus.karlsson@ivl.se*

In the late 1980s elevated levels of PCDD/Fs were discovered in sediments and biota from bodies of water receiving effluents from pulp mills. The forest industry responded to this potential environmental threat by eliminating elemental chlorine as a bleaching agent. This counter measure led to the virtual elimination of PCDD/Fs from the effluents, with concentrations falling below the detection level. However, as PCDD/Fs are resistant to degradation and have high affinity to particles, elevated levels may be found in the sediments of mill recipients long after the primary discharge has ceased. It is therefore relevant to question the extent to which sediment bound PCDD/Fs may propagate through aquatic ecosystem food webs and accumulate in, e.g. fish tissue.
In 2009-2010 a field sampling campaign was undertaken in the receiving waters of two Swedish bleached kraft pulp mills and reference areas located along the coast of the Baltic Sea at sampling sites previously investigated in the 1980s. Levels of PCDD/Fs were measured in water, sediment and various groups of biota including non-migratory fish. The overall pattern was that the levels of PCDD/Fs had decreased considerably compared to the situation in the 1980s both as the sum of individual PCDD/F congeners and after adjusting the values by the appropriate toxic equivalent factors (WHO-TEFs). However, there was a significant difference in the levels of PCDD/Fs in the sediments between the two mill recipients that may be explained by differences in morphometric characteristics of the systems.

In one of the studied coastal areas, where the bottom dynamic conditions favour the deposition and burial of fine matter, the levels of PCDD/Fs in the surface layers of the sediments were low (average 30 pg TEQ g-1 dw). A clear gradient was also found when analysing a sediment core; PCDD/F levels increased with sediment depth, indicating a declining trend over time. In the other mill recipient, surface sediment PCDD/F levels as high as 350 pg TEQ g-1 dw were recorded. This coastal area is shallow with an irregular bottom topography where sediment particles may undergo several resuspension cycles before finally coming to rest at greater depths in the open sea. The PCCD/F levels in muscle from non-migratory fish (perch) was also higher in the shallower recipient, 0.5 compared to 0.2 pg TEQ g-1 ww in the deeper soft bottom recipient and in the reference areas. For reference, it can be noted though that the recorded levels in fish from the same location and of the same, species, size, and age in the 1980s were above 20 pg TEQ g-1 ww. The results from this study highlight the need to take abiotic factors such as coastal morphometry into consideration when performing environmental risk assessments.

Keywords: PCDD/Fs, dioxins, Baltic Sea, pulp mill effluent, sediment, fish

8th International Conference on Fate and Effects of Pulp and Paper Mill Effluents

POSTER PRESENTATIONS

Point sources of impact on the Biobío River, Chile: Pulp and paper mill effluents and sewage treatment plant effluents, SIMILAR MAGNITUDE RESPONSES?

M. Saavedra[1], G. Chiang G.[1,2] and R. Barra[1].

[1]Environmental Science Center EULA – Chile, University of Concepción. P.O. Box 160- C. Concepción-Chile. E-mail: mariasaa@udec.cl; 2Canadian Rivers Institute and Department of Biology, University of New Brunswick, 100 Tucker Park Road, Saint John, NB, Canada

The Biobío river basin represents an important source of economic development. The productive sectors are forestry activities, along with metallurgical, chemical, oil refineries, textile, pulp and paper mill discharges and hydropower. This river provides multiple ecosystem services such as, drinking water, irrigation water, urban discharges receiver and industrial, power generation, recreation and fishing, it is extremely important as environmental relevance to support the various services it provides. The effects of exposure to pulp and paper mill discharges on fish have been studied since 2001 in the Biobío river basin, showing effects of endocrine disruption (increased levels of vitellogenin protein and gonadal maturation in sexually immature fish), increased levels MFO activity and physiological indexes (liversomatic and gonadosomatic index) and more recently in the Itata river with wild native fish populations showing effects of endocrine disruption (increase in 17β-estradiol and decrease of testosterone/11ketotestosterone gonadal production). The Biobío river not only receives pulp and paper mill discharges, there are other point sources discharges, such as sewage treatment plant effluent. Current results of juvenile individuals of O. mykiss exposed to different concentrations (0%, 12,5%, 25%, 50%, 75% and 100%) of treated sewage effluent, showed significant increase ($p< 0{,}05$) in the liversomatic index and induction of gonadal maturation, with 20% of oocytes in state III in sexually immature fish exposed to 100% effluent, ending the exposure time for 21 days. The results show similarities in the responses of biomarkers, but the responses of fish exposed to treated sewage effluent show lower values than fish exposed to pulp and paper mill discharges.

Keywords: Pulp and mill discharges, treated sewage effluents, fish, Biobio river, endocrine disruption

Ecotoxicity assessment of the industrial effluent from the Valdivia cellulose pulp mill and of its discharge receptor body (río Cruces)

E. Bay-Schmith[1], J. Silva[1], A. Cifuentes[1], C. Suárez[2] & R. Urrutia[2]

[1] *Laboratory of Bioassays, Dept. of Zoology, University of Concepción. P.O. Box 160- C. Concepción-Chile.*
[2] *Environmental Science Center EULA – Chile, University of Concepción. P.O. Box 160- C. Concepción-Chile.*

This research analyzes the ecotoxicological characteristic of the liquid discharges of one of the last pulp mill built in Chile: Planta Celulosa Valdivia. Its effluent falls into the Cruces River which in turn empties at the Carlos Anwandter Nature Sanctuary. During the year 2006 effluent and river water was evaluated for acute and chronic toxicity with six freshwater species: *Hyalella gracilicornis* (anphipod), *Daphnia obtusa* (water flea), *Oncorhynchus mykiss* (rainbow trout), *Gambusia affinis* (mosquito fish), *Selenastrum capricornutum* (microalgae) and *Lemna valdiviana* (duckweed).

Acute toxicity bioassays of effluents from processing pine or eucalyptus did not find lethal effects in any of exposed species. Chronic toxicity testing of the effluent did not show significant effects either. Only with *D. obtusa* certain level of chronic toxicity became evident after exposure of 21 days, with a fecundity reduction of 18 % (pine-effluent) and 32 % (eucalyptus-effluent) compared to control medium.

The results of the toxicity assessment to the river water revealed absence of lethal toxicity at the pre and post impact sites of the effluent discharge. No negative effect on growth or reproduction of the tested species was registered, except with *D. obtusa*. This species revealed significant differences in reproductive output in relation to the control when exposed to pre-impact water. This implies that contaminants up the river might have been polluting the water or, otherwise, natural variations in water quality might have affected *Daphnia's* fecundity at that time. Modern pulp production combined with appropriate waste treatment has improved considerably the environmental impact of chemical wood pulping.

Keywords: Wood pulping. Pine. Eucalyptus. Toxicity. Río Cruces

FATE MODELING / PROGRESS IN BIOACTIVE CHEMICALS CHARACTERIZATION

The development of a plume dispersion model for process water discharges

J. Malm, M. Malmaeus and M. Karlsson.

[1]*IVL Swedish Environmental Research Institute, P.O. Box 210 60, SE-100 31, Stockholm, Sweden. E-mail: magnus.karlsson@ivl.se*

According to Albert Einstein, "Everything should be as simple as possible, but not simpler." The key to successful environmental modelling lies in identifying the important processes while omitting the nonessential factors. Dispersion in aquatic environments is a complex process dependent on the sum of different movements created by, e.g. turbulent eddies, waves and currents. However, most observations and theories support the idea that the surface area of a given substance within a fluid increase proportionally to time squared. This behaviour was the fundament for a newly developed dispersion model for effluents in water recipients that can be applied to predict the initial dilution of waste water and cooling water discharges from, e.g. forest industries.

The main numerical method can be characterized as a cell-centred finite volume method for the conservation of mass and momentum. In the basic configuration of the model, the waste field is analysed using a polar discretisation and a spatial mean-value for dilution is computed in each segment in a steady-state solution. The analysis of temperature variation in waste fields due to thermal discharge can also be carried out by the model. It's also possible to use crude spatial boundary conditions for cases such as discharge into rivers.

The model has been applied in a dozen case studies of mill and power plant effluent dispersion in Swedish and Chilean coastal and inland waters. The results indicate that the model has the necessary qualifications to be an effective tool for environmental management.

Keywords: Plume dispersion, effluents, modelling, coastal and inland waters, impact assessment

INVERTEBRATE / FISH RESPONSES

Qualitative characteristics of blood cells and evaluation of hematological parameters in Trichomycterus areolatus (Valenciennes, 1846) and Percilia gillissi (Girad, 1854) exposed to pulp mill effluent in the Itata River (Chile)

A. Araneda[1], G. Chiang[1,2], A. Valenzuela[3], R. Barra[1]

[1]Environmental Sciences Centre EULA – Chile, University of Concepción. P.O. Box 160- C. Concepción-Chile. E-mail: anaraned@udec.cl;[2]Canadian Rivers Institute and Department of Biology, University of New Brunswick, 100 Tucker Park Road, Saint John, NB, Canada.; [3] Fish Laboratory, Faculty of Natural Science and Oceanography, University of Concepción, P.O. Box 160-C, Concepción, Chile.

Ichthyohematology is a great tool to assess the physiological changes, as we could use the study of morphological, biochemical and functional differences of the several cell lines, as well as the hematopoietic organs and the diseases related to those responses.

The study analyzed through relative count and morphometric characterization, the three blood lines, erythrocytes, leukocytes and trombocytes in two wild freshwater fish species populations (Trichomycterus areolatus & Percilia gillisssi), collected during summer 2009 in four sites upstream and three sites downstream a pulp mill effluent discharge in the Itata River (Biobío Region, Chile). As part of the study we calculated and assessed the fish condition factor, as general health status condition index.

The results showed that the morphological characteristics of the blood lines of T. areolatus y P. gillissi are equivalent to the described characteristics of other teleosts. Condition Factor and the immature erythrocytes showed a significative reduction in the first site downstream the pulp mill effluent discharge for both species. On the other hand, significative increment on leukocytes and trombocytes were observed, showing possible immunological alterations.

The hematological assessment for native fish shows no major differences in morphological characteristics from other teleosts and observed results in the three blood lines might be evidence of a possible anemia, along with immune alterations that might affect the condition factor downstream of the pulp mill effluent discharge, resulting in a stressed wild fish population.

Keywords: Pulp mill effluent, native fish, Ichthyohematology, Itata river

Pulp effluent effect on benthic macroinvertebrates community of central Chile

M. Díaz, C. Verdugo, K. Ovalle, F. Correa and R. Figueroa

[1]*Aquatic Systems Research Unit, Environmental Science Center EULA-Chile, University of Concepción, Box 160-C, Concepción, Chile. Email: mariaelisdiaz@udec.cl.*

The freshwater aquatic ecosystems are increasingly modified by several different human activities, being relevant industrial effluent discharges, which cause physical and chemical changes in the water column and sediment that result in structural changes in biological communities. Macroinvertebrate communities associated with river water bodies are considered bioindicators of environmental conditions, because they respond to different dynamics of natural and anthropogenic origin. Among the responses to such disturbance are movements of organisms into new areas allow drift in the stream (derived active) or are forced into this process (derived differential). The first is a mechanism that enables restoration of natural systems has been operated, because they are vectors of colonization processes. This behavior can be utilized as a tool to assessment a punctual impact, as drift occurs relatively even across the stretch of a river and offering optimal substrate for colonization, the possibility that they could be colonized, would be the same. This study evaluated the effect of a pulp effluent on macroinvertebrate colonization of artificial substrates (natural substrates previously treated). We compared three sampling areas defined as: pre-impact, impact and post impact (two stations at different distances) for two seasons. The registered macrofauna consisted, mainly immature stages of insects that showed significant differences between the impacted areas and non-impacted (ANOSIM $R= 0.51$, $p<0.001$), increasing the abundance of macroinvertebrates in the areas of impact and post-impact resistant for pollution families (e.g. Chironomidae, Lumbriculidae). The physico-chemical variables also showed significant differences (ANOSIM $R= 0.96$ $p<0.001$) being the conductivity, BOD5, dissolved oxygen variables that model the behavior of communities in the impacted areas in both seasons

Keywords: Pulp effluent, colonization, benthic macroinvertebrates

The environmental significance of forest industry effluents (pulp mills and papers) in the Biobio river basin

O. Parra, R. Figueroa, C. Valdovinos, E. Habit, H. Cid and C. Zaror
Centro de Ciencias Ambientales, EULA-Chile, Universidad de Concepción, oparra@udec.cl

The Biobío River basin (36°42'–38°49'S; 71°00'-73°20'O) is the third largest in surface (24,625 Km2), and the second longest river (380 Km), featuring the highest flow rate amongst Chilean rivers (average annual flow is 1160 m3/s). The Biobío river, shows increased stress due to impacts caused by human uses such as hydropower generation, water for irrigation, and pollution from urban and industrial wastewaters. Industrial effluents (three cellulose plants, two paper mills, and oil refinery), urban discharges and aquaculture wastewater represent the main point source of pollution, whereas agricultural activities account for most diffuse pollution. Since 1994, a thorough water quality monitoring program has been conducted, involving 24 physical-chemical parameters. Samples are taken at 8 points along the river's main course, and 6 additional registers taken at the inlet of principal tributaries. Samples are collected three times a year, to reflect seasonal conditions during spring, summer and winter. After collection, water samples were conserved, and transported according to Chilean standard procedures (INN-Chile, 1999). All analyses were conducted following Standard Methods (APHA-AWWA-WPCF, 2005). Results show that nutrients are present at very low concentration in the upper section of the Bio Bio river (BB1-BB3). However, the concentration of nitrogenated compounds – including ammonium, nitrite, nitrate, and total nitrogen- and total phosphorous significantly increase down the river, as a result of contributions from cellulose mills, urban settlements and agricultural/forestry activities. This situation is particularly critical in the lower section (between Negrete (BB4, middle part) and Concepción area (BB11-BB13 end part), where values close to eutrophication are found. Moreover, phenolic compounds also show a clear trend to increase from the upper to the lower part of river. However, total phenols concentrations along the river never exceeded 1 mg/L, which is considered an acceptable level. Furthermore halogenated compounds (AOXs), including pentachorophenol also show a tendency to increase downstream, especially from station BB4 due to cellulose and urban effluents. Other relevant water quality variables, like heavy metals, hydrocarbons and pesticides presented very low values or under the detection limits of methods. During the last decade, bleached kraft cellulose doubled its production in the Biobío river basin, reaching over 1.800.000 ton/year. Despite such drastic increase in production, water quality parameter directly associated to cellulose mills have not shown any variation respect to historical levels, showing the effect of new cleaner technologies used in such processes. A number of researches dealing with biomarkers and ecological status of aquatic biota have also been conducted, showing large-scale and long-term effect of human impacts on the river. Nevertheless, in terms of water quality, international criteria for protecting aquatic life are met, with the exception of ammonia, nitrite, and phenols in the lower part of the river. Thus significant ecological changes are the result of multiple stressors in the basin, including not only water quality changes, but important habitat structure and land use alterations throughout the watershed.

Keywords: Pulp mill effluent, native fish, Ichthyohematology, Itata river

Short term coastal environmental monitoring of a local pulp mill plant treatment outfall

P. Torres, R. Jerez, S. Figueroa and D. Arcos

Centro Regional de Estudios Ambientales, Universidad Católica de la Santísima Concepción. Avenida Cristóbal Colón # 2766, Interior. Talcahuano, Chile. E-mail: ptorres@ucsc.cl

Results of an environmental coastal program observation conducted off central Chilean coast off Constitución harbor are here discussed and related with a pulp mill treatment plant outfall within the open coast. The water type are characteristics of a coastal shallow area with a mix between fresh and salt water, especially during the winter times, with an presence of relatively cold water, salty and more dense at depth and high dissolved oxygen in the water column greater than 5 mlO2/l during the winter and relatively low level of dissolved oxygen less than 2 mlO2/l during the summer. The local water dynamics present a clear seasonal component parallel to the coast with a main NE component during the summer and a SW component during the winter.

The subtidal and intertidal fauna are clearly dominated by common benthic species described by several author for the central south of Chile. The census on marine bird and sea lions (O. flavescens) mammals both showed seasonal changes, with an increase of the local populations during the summer according with the reproductive behavior. Local presence of the Chilean dolphin (Cephalorhynchus eutropia) has also been associated with the mouth of the Maule River and other nearby coastal locations.

The chemistry of water quality on the entire coastal water column, on the sediment and marine organism have been describe as good and very good according with the water quality qualification. However, if we consider the variables of oil and grease, phenol and detergents the marine water quality occasionally can be described as bad and regular using the same qualification table (Conama - Chile, 2002). Important issue to considered in order to draw some conclusion is that the outfall of the pulp mill plant is closest to the Maule River mouth, where the non treated sewage from city of Constitución are discharged.

Consequently, the ecotoxicology analysis using standard bioassays methodology have not shown any lethal or sublethal effects on the water column off the pulp mill plan outfalls.

Keywords: Constitución- Chile, pulp mill plant , Programs of Environmental Alertness, O. flavescens, Cephalorhynchus eutropia, coastal environment

NUTRIENT ENRICHMENT

Monitoring of liquid waste pulp mill in a river water body

O. Parra, H. Cid and D. Avilés
Centro de Ciencias Ambientales, EULA-Chile, Universidad de Concepción, oparra@udec.cl

From the year 2004 began on the basis of Resolution Environmental Qualification (RCA No. 279/1998), the Environmental Monitoring of Arauco Valdivia Pulp Mill which incorporates among its a sub program its industrial effluent and the receiving water body, the Cruces River. The objective of this sub program is to monitor the influence of industrial discharge on physical, chemical and microbiological in the receiving body, the Cruces River, tributary of the Valdivia River basin (Lat Long W and W). Monitoring is carried out at three sampling stations, one located upstream of the discharge of industrial waste (E0 or control station), and two stations downstream of the spill, distant and 400 m and 800 m from the discharge point (diffuser) respectively. The methods of sampling and laboratory analysis correspond to those indicated by Standard Methods 21th Edition (2005). Currently, the monitoring program of the receiving body includes 55 parameters quarterly and 27 monthly, because in November 2009 (Res. Ex 108/09) were eliminated organochlorine pesticides and organophosphorus analysis. Therefore, the 90 parameters analyzed since 2005, is reduced to 55 from 2010, of which 28 are included in the guide for establishing CONAMA Guide for Secondary Water Quality Standards. To date, this program has been in operation phase of the Valdivia pulp mill 77 monthly samples. During this period, of the 28 parameters included in the Guide of CONAMA indicated above, only 5 parameters of the receiving body, stretch of river crossings, have not classified the water quality class of Exception, they are temperature, mercury, biochemical oxygen demand, suspended solids and fecal coliform. In the case of temperature, is because in some cases, the discharge makes a difference sometimes slightly greater than 1 ° C (from a Quality Class 1), in terms of mercury, it does not achieve the kind of exception because the method detection limit is above the limit set by the CONAMA Guide, BOD5 has only occasionally exceeded Exception class, and the same has happened with the total suspended solids. As for the Colimetry, in general, the control station has the same class as the downstream stations, so not for the influence of industrial effluent. It is also important to note that in addition to sampling of environmental parameters in concentration, is from the year 2010 the monthly analysis of the effluent charges which is consistent with the intent to regulate, based on effluent discharges to water bodies.
As a general conclusion, the conditions of water quality in the Río Cruces maintains the historical pattern and are generally very few parameters that are altered by the influence of industrial effluent and when this happens, they are expressed in a small way. In relation to the Guidelines for establishing CONAMA High Standards of Water Quality of reference, the Valdivia Plant effluent generates virtually no significant changes in the quality classes of the river system receiver, the Rio Cruces.

Keywords: Pulp mill effluent, native fish, Ichthyohematology, Itata river

PROTEOMICS, GENOMICS, METABOLOMICS

Validation of glucose-6-phosphate dehidrogenase activity as a biomarker of enviromental stress in larvae of Danio rerio (zebra fish).

E. Martínez[1], A. Llanos- Rivera [1,2], F. Cruzat[1,2] and R. González [1,2]

[1]*Unidad de Biotecnología Marina. Facultad de Ciencias Naturales y Oceanográficas, Universidad de Concepción.* [2]*Programa Investigación Marina de Excelencia (PIMEX). Universidad de Concepción. (fecruzat@udec.cl).*

The discharge of waste products of industrial activities to the aquatic environment may have an adverse effect on organisms. Among these effects, are alterations in metabolic pathways, which are identified with the use of biomarkers. In this context, this study proposed to validate both glucose-6-phosphate dehydrogenase (G6PDH) specific activity and mRNA expression, as biomarkers of environmental stress in the larvae of Danio rerio.

The experimental design involved static bioassays (144 hrs) where zebrafish embryos were incubated with increasing concentrations of the reference toxicant malathion. Upon completion of the bioassay, standardized as the point of complete yolk absorption, both the specific activity (by spectrophotometry) and the expression of G6PDH mRNA (by qRT-PCR) were quantified. The results indicated a significant increase in specific activity in larvae exposed to high concentrations of malathion ($\geq$ 8 ppm). At the same time, the mRNA expression levels increased at concentrations $\geq$ 3 ppm of the reference toxicant. Both results indicate that G6PDH is a suitable biomarker for assessing xenobiotic-induced alterations at the sub-lethal level. Furthermore, exposure bioassays with D. rerio larvae in the yolk absorption stage to a secondary effluent from CFI-Nueva Aldea, indicated a positive relationship between effluent concentration and the specific activity of the enzyme. However, the Tukey analysis indicated that only 100% effluent treatments produced statistically significant differences with the control.

These results indicate that the activity of this enzyme is a tool to for quantifying sub-lethal effects in the early stages of marine fish populations exposed to potential xenobiotics.

Keywords: Danio rerio, larvae, G6PDH, specific activity, expression, qRT-PCR

Support: PIMEX UdeC-Arauco
Acknowledgments: PFB-31/2007, COPAS-Sur Austral Program.

REPRODUCTIVE EFFECTS / ENDOCRINE DISRUPTION

A comparison of effluents and wood feedstocks from pulp and paper mills in Brazil, Canada and New Zealand and their ability to affect fish reproduction: Part II - chemical profiling and fish bioassays

R. Orrego[1], C. Milestone[2,3], J. Guchardi[1], P. Scott[2], A. Waye[4], M. Kohli[3], B. O'Connor[5], B. Smith[6], H. Engelhardt[6], D. McLatchy[2], V. Trudeau[4], T. Kovacs[5], T. Heid-Furley[7], A. Slade[8], M. Hewitt[3] and D. Holdway[1]

[1]University of Ontario Institute of Technology; [2]Wilfrid Laurier University, Waterloo, ON, Canada; [3]National Water Research Institute, Canada; [4]University of Ottawa, Ottawa, ON, Canada; [5]FPInnovations-Paprican, Pointe-Clare, QC, Canada; [6]University of Waterloo, Waterloo, ON, Canada; [7]Aplysia, Vitória, ES, Brazil;8Scion, Rotorua, New Zealand.

This study evaluated the ability of mill effluent extracts (solid phase extraction, SPE) from pulp producing countries (Canada, Brazil and New Zealand) to affect fish reproduction in a chronic exposure experiment using a natural population of immature rainbow trout (Oncorhynchus mykiss). Effluent samples were extracted in their country of origin and cartridges were shipped to Canada for bioassay and chemical evaluation The primary short-term reproductive endpoints included plasma vitellogenin (VTG) level and gonad somatic index (GSI) supplemented with molecular endpoints such as induction of hepatic ethoxyresorufin-O-deethylase (EROD) activity and changes in the activity of enzymes citrate synthase (CS) and lactate dehydrogenase (LDH) as indicators of liver respiratory-metabolic impairment. The exposure experimental protocol involved multiple intra-peritoneal injections of the dichloromethane SPE-DCM extracted material fraction (1 injection every 7 days for a total exposure period of 28 days), along with two steroid standards 17beta-estradiol (E2) and testosterone (TE). Initial analysis of variance/covariance (ANOVA/ANCOVA) demonstrated no differences associated with fish gender other than GSI (being male gonads being always smaller than female gonads). Moreover, no significant changes in GSI were observed during the experiment. Significant inductions of liver EROD enzymatic activities were observed in 1of 5 Brazilian mills, 1 of 4 Canadian mills and both New Zealand mill effluent extracts, in contrast to fish injected with steroid standards where no significant effects in EROD were observed. In addition, no significant differences were observed in CS and LDH enzymatic activities relative to the corresponding sampling time controls, suggesting that aerobic metabolic capacity in fish liver exposed to complex pulp and paper mill effluents extracts was unaffected and although some slight changes were observed they did not alter enzymatic detoxification activity nor the estrogenic effects observed. Significant increased levels of plasma VTG were observed in fish injected with E2 in 4 of 5 Brazilian and 1of 2 New Zealand mills, irrespective of the sampling time. Fish injected with TE did not show increased plasma VTG levels, and only a partial temporal increase of VTG levels was observed in all Canadian mills and the remaining Brazilian and New Zealand mills. Overall the results indicated an estrogenic effect of all tested mill extracts. Furthermore, a subsequent multi-variant Principal Component Analysis (PCA-NIPALS algorisms) has indicated that increased levels of plasma VTG were significantly related to industries using the Kraft process, using a mix of soft and hard wood, and using activated sludge treatment systems.

Keywords: Pulp mill, Medaka Embryo, d-rR, FLFII, teratogenesis.

Effect Potential of hormonally active substances contained in pulp effluent, over gonadal mature of male fishes of Trichomycterus areolatus and Percilia gillissi

R. Mendoza[1,3], P. Pezo[1], G. Chiang[3,4], R. Barra[3] and J. Gavilán[2]

[1]*Naturals Science and Oceanography, University of Concepción, P.O Box 160-C. Concepción-Chile. E-mail: rumendoza@udec.cl;* [2]*Biological Science, Celular Biology, Universidad de Concepción, Casilla 160-C, Concepción;* [3]*Environmental Science Center EULA – Chile, University of Concepción. P.O. Box 160- C. Concepción-Chile;* [4]*Canadian Rivers Institute and Department of Biology, University of New Brunswick, 100 Tucker Park Road, Saint John, NB, Canada.*

Chile presents 44 freshwater native species of fishes, this generates an important focus of study and conservation of their biodiversity. The Itata River present 12 native species and it was the object of pulp effluent discharge, which are composed of one complex mixture of chemical substance, this can generate alterations in the aquatic organism near to this discharge. For this the object of this investigation it was determinate, the possible effects to reproductive level over fish male of two natives species Trichomycterus areolatus and Percilia gillissi, through a comparative field study in the Itata River, evaluating the reproductive response (gonadal mature) of fishes. They are exposed in areas affect for those effluent (impact and post-impact), in comparison to areas that are not affect for these (pre-impact). We made physiological analyzes (gonadosomatic index and condition factor) and histological analyzes, to determinate the lobule's number as well as cysts's number in his different states. As to physiological index we find statistic differences only for condition factor in Trichomycterus areolatus, reflected in an increase in this in the impact zone. On the other hand both species showed no statistically significant differences to gonadal mature, however Percilia gillissi showed an increase on frecuency of cysts in spermatids and spermatocytes, for the impact zone with significant differences (Kruskal-Wallis, $p<0.05$), while Trichomycterus areolatus showed biological trends but no statistical support, which suggest a delay in gonadal maturation.

Keywords: Freshwater native species, pulp effluent discharge, effects to level reproductive, comparative field study, gonadal maturation, impact zone.

UNIVERSITY OF CONCEPCION MAP

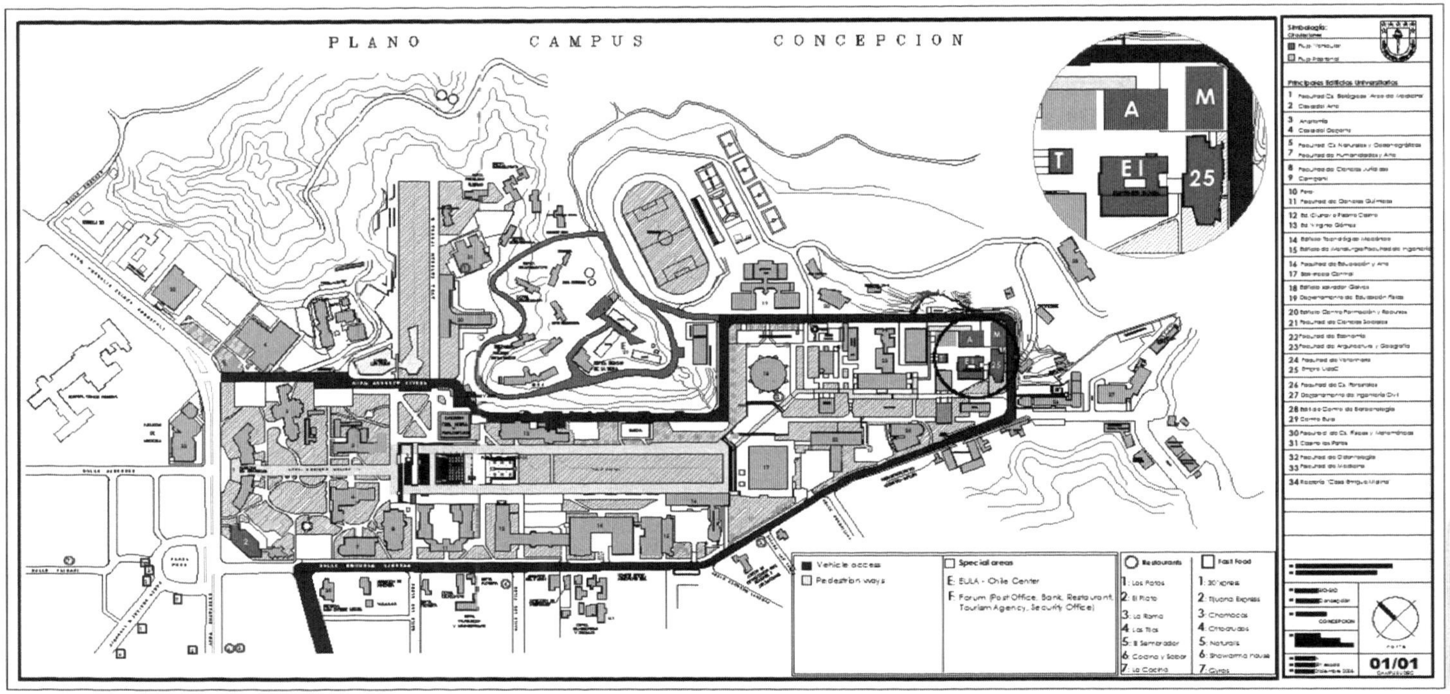

Sitios a utilizar en conferencia :

- 2 Casa del arte
- T Toilets
- A Rooms 1-4 - Registration and Information Desk
- EI Emprendo
- M Multiuse Outdoor Hall
- 25 Empre UdeC

ORGANIZED BY

Universidad de Concepción CHILE

IWA International Water Association

SETAC

PIMEX

UDT

CB UdeC

SUPPORT

ATCP Chile

CPL Consejo Nacional de Producción Limpia

CORMA

NBC Núcleo Biotecnología Curauma

UNESCO

Biobío

Gobierno de Chile

Ministerio del Medio Ambiente

Ministerio de Educación

Ministerio de Agricultura

Ministerio de Relaciones Exteriores

SERNATUR

SPONSORS

ARAUCO Sembremos Futuro

cmpc CELULOSA

Essbio

biodiversa

AQUALOGY

EcoRiles

amec

CHILE 2012

8-11 - Enero - Concepción

www.forestwater.cl